PIMLICO

637

PANDORA'S BREECHES

Patricia Fara is a Fellow of Clare College, Cambridge, and lectures in the History and Philosophy of Science Department. She is the author of several highly acclaimed books, including *Newton: The Making of Genius*, *An Entertainment for Angels: Electricity and Enlightenment* and *Sex, Botany and Empire: The Story of Carl Linnaeus and Joseph Banks*.

PANDORA'S BREECHES

Women, Science and Power
in the Enlightenment

PATRICIA FARA

PIMLICO

Published by Pimlico 2004

4 6 8 10 9 7 5

First published in Great Britain by
Pimlico 2004

Pimlico
Random House, 20 Vauxhall Bridge Road,
London SW1V 2SA

Random House Australia (Pty) Limited
20 Alfred Street, Milsons Point, Sydney,
New South Wales 2061, Australia

Random House New Zealand Limited
18 Poland Road, Glenfield,
Auckland 10, New Zealand

Random House South Africa (Pty) Limited
Isle of Houghton
Corner of Boundary Road & Carse O'Gowrie
Houghton 2198, South Africa

The Random House Group Limited Reg. No. 954009
www.randomhouse.co.uk

A CIP catalogue record for this book
is available from the British Library

ISBN 978-1-8441-3082-5 (from Jan '07)
ISBN 1-8441-3082-7

Papers used by The Random House Group Limited are natural,
recyclable products made from wood grown in sustainable forests;
the manufacturing processes conform to the environmental
regulations of the country of origin

Printed and bound in Great Britain by
Clays Ltd, St Ives plc

For Jim Secord

Contents

List of Illustrations

Acknowledgements

When I started writing this book, I realized for the first time how long I had been thinking about its major themes. So I should like to thank in retrospect many teachers, friends and colleagues both inside and outside the academic world who have helped me to consider these issues. More specifically, I am especially grateful to Jim Secord for his invaluable comments on a first draft, and also to Michael Fara, Stephen Gaukroger, Michael Hoskin, Sarah Hutton, Rebecca Stott and Judith Zinsser for their constructive criticisms of individual chapters. In addition, many thanks to my copy editor Mandy Greenfield, and also to my agent David Godwin and my editor at Pimlico, Will Sulkin, without whom this book would never have been published.

Pandora/Eve/Minerva

Prologue

Had God intended Women merely as a finer sort of Cattle, he would not have made them reasonable.

> Bathsua Makin, *An essay to revive the ancient education of gentlewomen*, 1673

Had Guy Fawkes struck again? When smoke billowed through the lobby of the House of Commons on 9 May 1792, some smouldering breeches stuffed with straw were found crammed above a lavatory ceiling. Suspicion immediately fell on Thomas Paine, the well-known revolutionary who had just completed his (metaphorically) incendiary *Rights of Man*. 'Guy Vaux' caricatures sneered at Paine's calls for equality, and opponents circulated small political tokens showing Paine hanging from a gibbet on one face and a pair of exploding trousers on the other (Figure 1). Sardonic mottoes reinforced the crude sketches: END OF PAIN and PANDORAS BREECHES.

In Greek mythology, when Pandora opened her giant casket, a host of evils flooded out across the world. Pandora's box was a common image for the French Revolution, and these copper coins warned against the radical activism that threatened to seep across the Channel into Britain. Several different versions were minted, coupling other suspicious characters with PANDORAS BREECHES to signal the dangers of political change.[1]

Pandora's breeches were a powerful symbol because the notion of a woman wearing trousers was outrageously funny, yet also frighteningly possible. The same year, the writer Mary Wollstonecraft published her own inflammatory book with a similar title to Paine's: *A Vindication of the*

Fig. 1
Pandora's breeches.
Copper coin, inscribed PANDORAS BREECHES/END OF PAIN, 1792.

Rights of Woman, often said to be the first manifesto for female emancipation. Calling for 'a REVOLUTION in female manners', Wollstonecraft insisted that women should use their minds. 'JUSTICE for one-half of the human race' meant giving women a better education; if men continued to oppress women by keeping them ignorant, she warned, then they would rebel – and a Pandora's 'box of mischief' would open up and destroy society.[2]

Unsurprisingly, conservative critics were appalled at Wollstonecraft's recommendations. Horrified by the consequences of losing their supremacy, men wrote venomous reviews that stressed the risks of giving women more freedom:

> For Mary verily would wear the breeches
> God help poor silly men from such usurping b——s.[3]

Amongst other insults, Wollstonecraft was vilified as a 'philosophising serpent'. Like the writhing snake below Pandora's breeches on the tokens, this was a reference to Eve's corrupting influence on Adam when she tempted him to bite into the forbidden apple. '*Pandora*'s Box,' explained the standard text on Greek myths at that time, 'may properly be took in the same mystical Sense with the Apple in the Book of *Genesis*.' In the Bible, Eve – the first woman – is the source of sin, held responsible for

4

human banishment from the Garden of Eden. In classical mythology, Pandora was the first woman, similarly blamed for the spread of wickedness and the end of the Golden Age. Eve or Pandora, the moral of the story is the same. Throughout the Western world, women have been reviled as the origin of evil: art and literature are saturated with metaphorical as well as literal references to women as temptresses, seducers, witches.[4]

Wollstonecraft railed against these tales of lost innocence. 'We must get entirely clear of all the notions drawn from the wild traditions of original sin,' she protested; 'the eating of the apple . . . the opening of Pandora's box, and the other fables, too tedious to enumerate, on which priests have erected their tremendous structures of imposition . . .'[5] She campaigned to reform the traditional constraints that prevented women from achieving their full potential. Sympathisers associated Wollstonecraft not with Eve or Pandora, but with a third symbolic woman – Minerva, the goddess of wisdom. Eve and Pandora represented the lust for forbidden knowledge, but Minerva stood for learning. During the Renaissance, Minerva – the Roman equivalent of Athene, patron of Athens – became a potent symbol of scholarship.

The concept of an intellectual woman had long been riddled with ambiguities. Some of these are alluded to in Figure 2, an advertisement for an early-nineteenth-century encyclopaedia. According to classical mythology, Minerva had sprung fully armed from the head of Jupiter. So is she a warrior or a scholar? Does this abnormal birth signify that women are incapable of passing on intellectual gifts? Here Minerva clasps a spear in her strong right arm and a soldier's helmet provides a perch for her learned owl, the wise bird that can see in the dark as well as the light. At her feet, her military shield is adorned by Medusa, the female Gorgon whom Minerva had helped to slay and whose petrifying appearance froze ignorant men to stone.

Mathematics, history, astronomy: like knowledge itself, all the academic subjects were symbolised by female figures, but studied by men. The swept-back curtains reveal the world of learning, the source of the wisdom symbolised by Minerva. Interrupted in his geometrical calculations, a dedicated student – male, inevitably – lets fall his dividers to heed Minerva's advice. Learning was the preserve of men, and educated women challenged the very definitions of sexual difference. How could they truly be women if they displayed a gift for rational thought? Wollstonecraft

Fig. 2
Minerva Directing Study to the Attainment of Universal Knowledge.
Frontispiece of the *New Encyclopædia*, 1817.

complained that she had 'been led to imagine that the few extraordinary women who have rushed in eccentrical directions out of the orbit prescribed to their sex, were *male* spirits, confined by mistake in female frames.'[6] Relying on unfeminine astronomical imagery, even the very phrasing of her protest transgressed convention.

6

One man who did admire Wollstonecraft was James Barry, an eminent artist who specialised in allegorical history paintings. Wollstonecraft's eloquent books, he wrote, demonstrated why the ancients had made Minerva a woman. For thirty years, Barry struggled to understand a paradox of his own time: how could Minerva be a woman, when learning was a man's activity? Women who engaged in intellectual activities contradicted all the social norms.

Time and again, Barry returned to what he regarded as his most important picture – the first part of Pandora's myth, the story of her creation (Figure 3). On Jupiter's instructions, Pandora was modelled out of clay, and then presented to the Olympian deities so that she could learn their skills. In Barry's final version of her birth, the curvaceous Pandora reclines on a chair, surrounded by Venus and three delicate Graces who are carefully dressing her. Pandora displays the fashionable ideal of feminine beauty – her shape and pose make a strong contrast with the triumvirate of male gods seated to the right of the picture. Barry remained preoccupied with this scene, experimenting with ways of portraying 'Pandora, or the

Fig. 3
The Birth of Pandora.
James Barry, etching and engraving, *c.* 1804–5.

Heathen Eve, brought into the assembly of the gods, attired by Venus and the Graces, and instructed in the domestic duties of a wife by Minerva'.[7]

'The domestic duties of a wife': Minerva looms over Pandora, handing her a shuttle to represent the female art of weaving. Standing at the very centre of the picture, Minerva is a strangely ambiguous figure. Although she has breasts, she wears a military helmet, and her muscular arms and stalwart posture make her resemble Jupiter and his colleagues rather than Pandora and her fragile attendants. Unlike the voluptuous and gentle Pandora-Eve, Minerva is a powerful goddess of masculine proportions.

Minerva was the icon of learning, yet here she seems to be teaching Pandora non-intellectual skills of the sort reserved for women. Scholarly women were disturbing: no accident that Minerva had such a manly body. For one thing, men worried about their own well-being. As Samuel Johnson pointed out, 'A man is in general better pleased when he has a good dinner upon his table, than when his wife talks Greek.'[8] There were two major advantages of letting women study: they would become more interesting companions for their husbands, and they would be able to educate their sons. Benjamin Franklin wondered what other reason a young woman could possibly have for learning about science. 'But why will you,' he asked a would-be pupil, 'make yourself still more amiable, and a more desirable Companion for a Man of Understanding, when you are determin'd, as I hear, to live Single?'[9]

Popular journals predicted the disruption that would ensue if a woman devoted herself to physics: 'While she was contemplating the Regularity of the Motions of the heavenly Bodies, very irregular would be the Proceedings of her Children and Servants; the more she saw of Order and Harmony above, the more Confusion and Disorder would she occasion in her domestick Affairs below; the more abstracted she was in her Ideas and Speculations, the greater Stranger would she be to the Rules and maxims of common Prudence.'[10] If women learned to wear the breeches, then a veritable Pandora's casket of troubles would be unleashed.

Chapter 1

Women / Science

What cruel Laws depress the female Kind,
To humble Cares and servile Tasks confin'd? . . .
That haughty Man, unrival'd and alone,
May boast the World of Science all his own:
As barb'rous Tyrants, to secure their Sway,
Conclude that Ignorance will best obey.

Elizabeth Tollet, 'Hypatia', 1724

In *A Room of One's Own*, Virginia Woolf wondered what would have happened if William Shakespeare had had an equally gifted sister. Empathetically, she envisaged her imaginary Judith following William's example and running away to London, only to meet a very different destiny – mockery, pregnancy and a lonely suicide. Woolf explained that Judith was doubly shackled. Most obviously, she lacked her brother's education, having been taught domestic skills to attract a wealthy husband while William went off to school. More insidiously, Judith had been conditioned from birth into accepting the confining norms of sixteenth-century society, so that the very act of trying to break free would have driven her mad.[1]

But suppose Newton or Descartes or Darwin had had a clever sister. What would be the fate of a scientific Judith who tried the equivalent of running away to London and knocking at the stage door? Perhaps, as Woolf conjectured, frustrated female geniuses became lonely, half-crazed recluses, mocked and feared as witches with extraordinary powers. But intelligent women could find ways to accommodate their intellectual interests within conventional lives. In wealthy families, some exceptional women employed impoverished scholars for private tuition. More commonly and more importantly, well into the nineteenth century most scientific activity took place in private homes, not in large laboratories or research institutions. This meant that although women were excluded from universities and academic societies, they did become involved in science.[2]

Most typically, women became engaged at a practical level when a male relative was carrying out research. Their domestic responsibilities already included managing the household, producing and caring for children, and providing the emotional and physical care that liberated their men from daily chores. Exhibiting varying degrees of meekness, they accepted being also co-opted as menial assistants who received no acknowledgement for their work. Using the organisational skills that they had acquired to run the household, sisters, wives and daughters administered the family's experimental investigations – employing and supervising assistants, buying materials and equipment, marketing medicines and instruments to pay the bills.

Bathsua Makin, a seventeenth-century campaigner for educational reform, pointed out that a housewife's work demanded expertise in areas that we would now call scientific: 'To buy Wooll and Flax, to die Scarlet and Purple, requires skill in Natural Philosophy. To consider a Field, the quantity and quality, requires knowledge in Geometry. To plant a Vineyard, requires understanding in Husbandry: She could not Merchandise, without knowledge in Arithmetick: She could not Govern so great a Family well, without knowledge in Politicks and Oeconomicks: She could not look well to the wayes of her Houshold, except she understood Physick and Chirurgery.'[3]

Wives, sisters and patrons played indispensable parts in achieving the results for which their men became renowned. Some women taught themselves foreign languages, so that they could keep their husbands or brothers up-to-date with the latest scientific results from abroad. Others suggested new theoretical interpretations, collected botanical and geological specimens, or set up classification schemes. Women were often responsible for collating, editing, illustrating and publishing books that inevitably appeared under the male partner's name. They also undertook the vital tasks of translating and interpreting complicated ideas. By writing lucid explanations, they ensured that scientific knowledge became accessible to everyone – future scientists as well as the general public.

Delving back through the centuries, feminist historians have rewritten women's lives according to modern standards of equality and liberation. This anachronistic approach can involve much wishful thinking, because these women behaved and wrote in ways that simply do not conform with today's ideals of independence and equality. Many women of the past seem almost to be colluding in their own oppression, themselves accepting that they could not fulfil the same functions as their husbands, sons and brothers. Even those who did rail against their limited opportunities generally believed that, physically and psychologically, they were suited to different work from men.

Aristotle's ideas about bodies and minds prevailed well into the seventeenth century and even beyond. Women were regarded as inferior versions of men, placed beneath them on the great chain of being that stretched from the lowest organisms up towards the angels and God. Men were like the sun – hot and dry. But women resembled the cold, moist moon, so that their brains were less capable of rational thought. Long after these ancient models had been overturned, new physical criteria – anatomical differences, hormonal systems – provided new rationales for keeping women below men in the intellectual hierarchy.[4]

Time after time, people interested in maintaining the *status quo* spelt out the appropriate roles for men and women. In an eighteenth-century stage comedy, Lady Science confesses, 'I am justly made a Fool of, for aiming to be a Philosopher – I ought to suffer like *Phæton*, for affecting to move into a *Sphere* that did not belong to me.' The dastardly Gainlove rejects Lady Science and marries her daughter instead, because, he insists, 'the Dressing-Room, not the Study, is the Lady's Province – and a Woman makes as ridiculous a Figure, poring over Globes, or thro' a Telescope, as a Man would with a Pair of *Preservers* mending Lace'.[5]

Lady Science and her erstwhile suitor bisected society into separate spheres: men studying science, women learning domestic skills; men exploring the outside world, women confined to hearth and home. This simplistic model – separate spheres – dominates how we perceive life in earlier centuries. In another satirical play about learned women, an angry father can hardly tolerate that he has 'a Daughter run mad after Philosophy,

I'll ne'er suffer it in the Rage I am in; I'll throw all the Books and Mathematical Instruments out of the Window'.[6] This was, of course, a hollow threat – the audience was not expected to believe that he would really treat her things that way. Rhetorical claims are often exaggerated, and there was no shortage

Fig. 4
Margaret Cavendish (Duchess of Newcastle) and her family.
Frontispiece of Margaret Cavendish, *Natures Pictures Drawn by Fancies Pencil
to the Life* (London, for J. Martin and J. Allestyre, 1656); original by
Abraham van Diepenbeke, engraved by Peter Clouwet.

of frightened but dogmatic protests that women were incapable of engaging in such an unsuitable activity as natural philosophy.

This propaganda campaign still distorts how we view the past. But in fact women strongly affected the development of scientific ideas and their integration within society. If we peer behind the front door of scientific households, a different picture appears. Figure 4 shows the Newcastle family at dinner. Margaret Cavendish and her husband the Duke of Newcastle sit at the right, their heads crowned with laurel wreaths. This picture was doctored for publication. Here the Duke of Newcastle is speaking, while his wife listens mutely at his side. In the original version, Cavendish was holding up her hand for attention as she addressed the party. In private settings, women enjoyed far more power than in the outside world.[7]

In the middle of the seventeenth century, the Newcastle couple formed the nucleus of an important intellectual circle. Both strong royalists, their moves between Paris and London depended on the English political situation. They invited many famous scholars, including René Descartes and Thomas Hobbes, to share their meals and discuss the latest philosophical and scientific ideas. Unlike now, intellectual debates took place in people's homes rather than in official institutions such as universities or societies. Margaret Cavendish could participate in the learned conversations that were taking place in the privacy of her dining-room, even though she did not always have the courage to venture her own opinions.

Cavendish dispensed useful advice to women who were frustrated by their conventional education: choose the right man. She had married into one of England's richest families. In Figure 4, the fine wood panelling and armorial crest over the mantelpiece advertise that she lived in stately surroundings; a servant is opening the window to counteract the heat from the large fire. Her husband, like many seventeenth-century Cambridge graduates, knew far more about the rules of hunting than the laws of nature, but he enjoyed supporting gifted scholars. He encouraged his wife's intellectual interests, financed the publication of her books, and conveniently had a well-educated brother who taught her natural philosophy.[8]

Some scientific women were born into rich families; others, like Cavendish, chose husbands whose interests meshed with their own, so that they could study natural philosophy and collaborate in experimental investigations. Unlike men, these women only rarely left published texts behind, but their letters and personal notebooks demonstrate how actively they engaged in scientific and philosophical controversies of the time.[9] Katherine Jones (Lady Ranelagh) lived with her brother, the famous chemist Robert Boyle, for thirty years. She shared her brother's laboratory equipment and engaged in debates with their visitors, both men and women. She belonged to an intellectual network of wealthy women who experimented and wrote inside their own homes, women like Mary Evelyn who experimented alongside her husband John, the famous diarist, and gradually took over the record-keeping.[10]

These wealthy female aristocrats were generally banned from venturing out into the academic world, but some of them converted their homes and stately mansions into discussion centres. Here they could debate with Europe's most distinguished intellectuals and participate in the extensive correspondence network that bonded together the Republic of Letters, the international intellectual community which transcended geographical boundaries. Some of them employed scholars as teachers or invited them as long-stay house guests; this meant that learned protégés relied on women's patronage for survival. John Aubrey reported that the wealthy Countess of Pembroke made sure 'Wilton house was like a College, there were so many learned and ingeniose persons. She was the greatest patronesse of witt and learning of any lady in her time. She was a great chemist and spent yearly a great deale in that study. She kept for her laborator in the house Adrian Gilbert . . . She also gave an honourable yearly pension to Dr Mouffett, who hath writt a booke *De insectis*. Also one Boston, a good chymist . . .'[11]

Just like their brothers and husbands, wealthy women could choose to support scholars financially. To return the debt, authors paid their patrons lavish compliments in printed books, and thus advertised this female contribution to the sciences. In the dedication for his book on comets, the musician Charles Burney wrote an obsequious tribute to the Countess of Pembroke (a later one than John Aubrey's) praising her enthusiasm for astronomy. Another strategy was to publish texts in the form of letters between a learned natural philosopher and a distinguished woman. Leonhard Euler's two volumes of simplified science were addressed to a German princess and sold exceptionally well throughout Europe; for his imaginary correspondent, the Swiss geological historian Jean de Luc chose his English patron, Queen Charlotte.

During the eighteenth century, it became more common for intellectual women to gather together in discussion groups, such as the bluestockings in London. Especially in France, women hosted weekly meetings that were also attended by men, and where the latest scientific results were often discussed along with new literature and political scandals. Although many of these women have been forgotten, they exerted considerable power. Elizabeth Ferrand is now unknown, yet her *salon* attracted Denis Diderot and other eminent Parisian men of letters, such as the well-known philosopher Étienne Bonnot de Condillac. According to malicious gossips, Ferrand was a dour humourless woman, yet they all paid tribute to her mathematical ability. Art connoisseurs singled out for praise her portrait, which was first displayed in the 1753 Louvre exhibition and pays tribute to her scientific expertise – momentarily distracted from her studies, she gazes out at the viewer, a large volume on Newtonian physics propped up like a Bible beside her. Condillac did have the grace to acknowledge that her ideas lay at the heart of his *Treatise on the Sensations* – but only his name appears on the title-page.[12]

Other wealthy women indulged their fascination with science by building up large collections of shells, minerals and pressed flowers. Sarah Sophia Banks was a collecting addict, and she possessed the same passion for natural history as her brother Joseph, who was President of the Royal Society for more than forty years. Perhaps they both inherited their enthusiasm from their mother Sarah – the botanical herbal she kept in her dressing-room was the first serious book ever read by Joseph, who was not a brilliant scholar. His younger sister Sarah Sophia, on the other hand, seems to have been a frustrated academic. As an adult, she was ridiculed for stuffing her pockets with books so that she would never be short of something to read. Had she been a man, her inelegant clothes and studious demeanour would have been praised as signs of her intellectual aptitude. Instead, she was mocked for lacking the appropriate feminine graces. Joseph Banks did marry, but he seems to have attached more importance to Sarah Sophia, who lived with the couple until she died. The brother and sister formed an inseparable scientific partnership. Although she could not participate in his life at the Royal Society and London's gentlemanly clubs, she was intimately involved in the scientific meetings and discussions that took place in their central Soho home.[13]

These women's activities were not just idiosyncratic pastimes – building up large collections of samples from all over the world was vital for classifying the exploding number of discoveries. The Duchess of Portland (as a later Margaret Cavendish is usually called) started her enormous collection of plants, minerals and fossils when James Cook gave her some shells after one of his voyages to Australia. The auction after her death lasted for thirty-eight days, when the curiosities she had accumulated were divided up between eminent bidders, including the surgeon William Hunter. Another important yet forgotten woman collector was Madame Dubois-Jourdain – even her Christian name seems to have disappeared from the records as her identity became subsumed into her husband's. Her husband was attached to the French royal household, and Dubois-Jourdain studied physics and chemistry so that she could catalogue her large collection of minerals. She became linked in to an extensive correspondence network of mineralogists, and enthusiasts came to admire her cabinet in her Paris home so that they could exchange and discuss rare rocks with her. By the time that Dubois-Jourdain died, she had acquired many specimens that were scientifically as well as financially valuable.[14]

But these were the privileged few. Scholars have been able to reconstruct their lives – if only partially – by examining family records, diaries and letters. All over Europe, similar conversational scenes to Cavendish's were taking place in less wealthy households, although it is even more difficult to learn about these women's activities in any detail. Because experimental natural philosophy was a relatively new activity, men who worked at home not only relied on their womenfolk's cooperation, but also had to negotiate ways of integrating scientific projects within the daily household routine. Arguments about the latest scientific ideas were thrashed out around the dinner table, in mixed groups that gave women the opportunity to listen and perhaps contribute their own ideas. Women were actively involved in the family's experiments and could participate in discussions about their results.[15]

Very little information about these domestic situations has survived, so it is impossible to discover how many women collaborated in experimental research and took part in dinner-party debates. This female participation was undoubtedly far greater than old-fashioned accounts suggest. For one thing, experiments often require more than one participant: astronomers dictate their observations while looking through a telescope; chemists cannot simultaneously pour in reagents and record measurements; physiologists need several hands to hold down specimens and manipulate instruments.

In the seventeenth and eighteenth centuries, before being a scientist had become a professional career option, experimenters struggled to earn a living. The women in the household provided a free source of labour, and many brief – but solid – pieces of evidence testify to their involvement, such as passing references in letters or the occasional grateful tribute in a preface. Figure 5 is a rare illustration of a woman participating in an experiment. The French lecturer Jean Nollet earned money by performing demonstrations designed to enthral as well as instruct his audience. When electricity became the most fashionable scientific entertainment, Nollet advertised his new machine, which needed a helper to turn the handle. Rarely even mentioned, such assistants are normally unidentifiable, indicated only symbolically by a disembodied hand in a frilly white cuff.

Fig. 5
Jean Nollet and his female assistant.
J. A. Nollet, *Essai sur l'Électricité des Corps* (Paris, 1746).

However, Nollet's picture makes it clear that his invisible assistant was a woman (although she is not referred to in the text).

Scientific experimenters, lecturers and entrepreneurs lived and worked in the same place. Modern laboratories and factories are now completely separated from people's homes, but the whole family used to be involved in running businesses. When the engineer James Watt was away from home, he left his first wife Peggy in charge of their instrument trade, and later wrote technical letters to his second wife Annie, knowing that she had learned chemistry during her upbringing in a linen-bleaching family. Watt's friend William Withering, a famous botanist and physician, instructed his children to collect plants while they were out walking, and then got the French governess to draw them for him – although his daughter reported sabotaging science by painting patterns on fungi to fool him. Withering set himself up as an expert who taught science to women through correspondence, yet he owed his fame as a heart specialist to an elderly and uneducated woman's traditional herbal remedies.[16]

Many scientific experimenters expected the women around them to work. One of Newton's enemies, the Royal Astronomer John Flamsteed, recruited his wife Margaret to study alongside his paid apprentices and help him make observations. Her own notebook reveals her mathematical proficiency and the range of tasks she had to cope with. After her husband's death, Margaret Flamsteed superintended the publication of his star catalogue, which remained a major reference source for more than a hundred years.[17]

The French astronomer Joseph Lalande also relied on women mathematicians – including his own niece by marriage – and was especially close to Nicole-Reine Lepaute. She was married to the royal clock-maker, and had already computed detailed tables for his experiments on pendulums. Lalande enrolled Lepaute to help predict the precise date that Halley's comet would return in the mid-eighteenth century, and for months on end she sat at a table with Lalande carrying out long and complicated calculations. Lepaute was constantly engaged in astronomical mathematics for around twenty-five years, but – like other female calculators – her work was not always acknowledged by the men who depended on it. Lalande himself did pay tribute to female astronomers, but excused the neglect of his colleagues by blaming the jealousy of other women who could not bear to be outclassed.[18]

John Desaguliers, Newton's favoured assistant at the Royal Society, converted his home into a scientific school. Apparatus was stashed all over their Westminster house – instruments in various stages of completion, piles of unsold books, and bulky demonstration equipment including an eight-foot-wide bellows and a working steam-engine. When Desaguliers went on his frequent international lecturing tours, his wife Joanna was left behind to run the school as well as caring for the paying boarders. Small wonder that five of her seven children died in infancy – and who, one wonders, took over when he was laid up every winter with attacks of gout?[19]

As well as organising homes, children and businesses, women were involved in developing new techniques. When the factory owner Josiah Wedgwood tried out different methods of firing and colouring pottery, his wife Sally helped him record the results. He explained to his business partner how he enabled her to go on working with him, even though she was already overburdened with household affairs – including their three-month-old daughter: 'Sally is my chief helpmate in this as well as other things, and that she may not be hurried by having too many *Irons in the fire* as the phrase is I have ordered the spinning wheel into the Lumber room.'[20]

During his trials of different types of gases, the chemist Joseph Priestley made his wife Mary responsible for keeping his experimental mice warm on the mantelpiece. There were, of course, hitches. Mary was disconcerted to discover chemical flasks and minerals stowed away in her travelling trunk of neatly packed clothes, and Joseph was furious when their daughter carefully washed out all his laboratory bottles.[21] Tempting to laugh – but these intimate details of domestic disruption, so painfully familiar and imaginable, vividly convey the realities of early scientific research. Women are absent from the written reports, but in reality they were very much present.

Official accounts of Soviet Russia avoided mentioning Josef Stalin. In contrast, women have not been written out of the history of science: they have never been written in. This neglect is part of the large-scale omission of women from the historical record, but there is no simple way of rectifying the situation. Feminists have chosen several different routes to restore these vanished women to visibility. In the late 1970s, the artist Judy Chicago shocked conservative critics by exhibiting an imaginary dinner party between forgotten women from the past. She painted the floor with the names of 999 high-achieving women, and designed individual place settings around a triangular table, arranging her female guests in a chronological loop so that Virginia Woolf sat next to an ancient goddess.[22]

But celebrating people just because they are women suggests that they are bound together by some essential quality of womanhood which transcends the barriers of time and culture. How could a modern overworked doctor communicate with a refined Renaissance lady of leisure – how would she argue with a woman who believed that she possessed a cold, soft brain incapable of intellectual reasoning, and a womb that wandered round her body causing all sorts of mysterious disorders? How could a company director explain her professional pride to a wife whose whole upbringing had been dedicated towards preparation for marriage and financial security as she was transferred from father to husband? How could a university scientist discuss her research with a well-educated aristocrat who insisted that the Bible was the major source of knowledge about the universe, and that God had created the world at 9 am on 23 October 4004 BC?[23]

Too much sympathy for these neglected women can be counterproductive. Emphasising the difficulties faced by intelligent women can convert them into self-sacrificial martyrs. In well-intentioned pastiches of the past, scientific women emerge as cardboard cutouts – the selfless helpmate, the source of inspiration, the dedicated assistant who sacrifices everything for the sake of her man and the cause of science. On the other hand, overcompensation – glorifying women as lone pioneers, as unrecognised geniuses – also has its drawbacks. Despite having to struggle against huge disadvantages, such arguments run, some women did contribute to scientific progress. If only they had biologically been men, one can almost hear their biographers sigh, then their true brilliance would have been recognised. Prominent examples include Aspasia of Miletus, Hypatia of Alexandria and Hildegard of Bingen. All exceptional women, without doubt, but it is misleading to celebrate them as suppressed scientists. Modern science bears little resemblance to intellectual pursuits of ancient Greece, fifth-century Egypt or Benedictine monasteries. These women certainly deserve to be honoured, but only within the framework of their own contemporaries.[24]

There is no point in distorting women's importance by exaggerating their activities. Singling women out as geniuses is as misleading as suppressing their existence. Standard caricatures of women – the docile assistant, the doting but ignorant source of inspiration – are certainly demeaning descriptions. But substituting yet another stereotype – the lonely, unappreciated pioneer – gets no nearer to understanding women's status in science and the lives they led. Making women from the past into brilliant proto-scientists is just creating a female version of solitary male geniuses. More realistic models are needed for both the sexes.

Trying to squeeze women into conventional stereotypes makes it impossible to reconstruct how they felt about their own lives, what they themselves regarded as their significant achievements. Part of the problem is finding an appropriate style for telling women's lives. As the famous clerihew quips: 'Geography is about maps, / But Biography is about chaps.' There is no established format for integrating a woman's personal and professional experiences within the covers of a single book. Woolf complained that too much history is about wars, and too many biographies are about men. Her own father had narrated George Eliot's career as though this female novelist were a purely intellectual being: through selective pruning, he effectively eliminated her tumultuous emotional existence. Old-fashioned biographical conventions force women's lives to follow male scripts by emphasising publicly acclaimed successes, rather than inner feelings and personal relationships.[25]

This problem is even more acute in scientific biographies. Until recently, words like physicist, philosopher and mathematician automatically signified a man, so that scientific women were seen as freakish intruders into a male domain. Sympathetic biographers worry that including emotional aspects of a woman's life will reinforce all those existing prejudices that women do not belong in the cold world of scientific research. Men as well as women are affected by their family and financial worries, their loves and losses, joys and despairs, yet the ethos of science demands that its practitioners perform with icy detachment. This ideal clashes with stereotypes of sensitive nurturing women and female scientists have been made to feel pariahs twice over – not only as invaders into a hard masculine world, but also as outlaws excluded from cosy feminine circles. They were parodied as atypical oddities who had renounced all claims to normal female interests, yet were unable to participate fully in scientific life.

Take the case of Marie Curie. Her daughter Eve portrayed her as a saintly mother, an exemplary martyr who dedicated herself to her family as well as to science. Like many other biographers, Eve glossed over the scandal of Marie's affair with a married colleague. Female scientists are supposed to be asexual, abnormal women, and Marie Curie is always shown wearing plain clothes and a humourless expression. Marie and her husband Pierre were complementary scientific partners – she was a physicist and he was a chemist; she was outgoing and energetic, he was a nervous recluse whose behaviour would nowadays lead to a diagnosis of attention deficit disorder. Nevertheless, in the romanticised versions of their lives, it is Marie who appears as an obsessive chemist. Her success is attributed not to her intellectual originality but to a daily grind of methodically sifting through tons of pitchblende – the scientific equivalent of mundane cookery. Marie Curie has been converted into an icon for ambitious young women to emulate, yet this image of a badly dressed drudge warns female students that they must abandon all prospects of normality if they wish to compete in the scientific race.[26]

A better way of highlighting the significance of women in science is to tackle conventional history head-on and rewrite it. Science's history is about far more than equations, instruments and great men – it is about understanding how a huge range of practical as well as scholarly activities became the foundations of our scientific and technological society. Women played vital roles in that transformation.

There is no right way to create the past. After the Second World War, optimistic campaigners seized on science as a replacement for Christianity, as a secular religion that would unite the nations to bring peace and progress. They rewrote European history, placing the birth of modern society not in the artistic Renaissance, but in a seventeenth-century Scientific Revolution – a term that had only been introduced in the 1930s. For these scientific historians, science meant ideas: they were interested in abstract theories about gravity and the structure of the universe. They divorced science from daily events and world affairs, and studied the scholarly debates between leisured academics and clerics. Concentrating on physics and astronomy, they told science as an epic success story, a triumphant march towards incontrovertible truth led by great heroes such as Galileo, Kepler and Newton.

Subsequent generations have brought new problems, new questions, new ways of writing science's history. The Scientific Revolution, seen as so important for decades, is being ironed out of existence. Historians now emphasise continuity rather than change. They are challenging the story that science erupted suddenly in early modern Europe and are replacing it with different versions of the past. Many experts see a gradual transition around the turn of the eighteenth and nineteenth centuries, when modern scientific disciplines were being created, the financial rewards of research were being recognised, and research was starting to move out of people's private homes into laboratories, museums and hospitals. After all, it was only in the early nineteenth century (1833, to be precise) that the word 'scientist' was invented – a new term to identify a new social category. Before then, investigations that we would call scientific were carried out by skilled artisans and by a vanished group labelled natural philosophers, who were mostly wealthy gentlemen trying to learn more about God by studying the natural world.

Instead of focusing exclusively on great minds and great ideas, historians are now more interested in examining how science has entered everyday life. They are investigating the activities of people who did play vital roles in science's history, but who worked outside learned societies or universities. These unrecognised experts included skilled artisans who made surveying instruments or calculated the trajectories of cannon balls, and practical people – navigators, herbalists, miners – who possessed valuable stores of knowledge unknown to closeted scholars. In romanticised versions of the past, science progresses in uneven leaps as solitary geniuses make momentous discoveries in their disinterested search for truth. Other accounts – aiming to be more realistic – recognise that scientific researchers are human beings prompted by mundane concerns, including money and fame as well as industrial, medical and military requirements. And some writers have introduced women into science's past.[27]

In conventional versions of science's history, women are either absent, or else feature as useful appendages of a famous man – the admiring wife, the helpful sister, the docile pupil. This is partly because stories about science have been written like schoolboy adventure novels. Bristling with the vocabulary of warfare and competitive sport, they feature scholarly gladiators triumphantly battling against the forces of nature. Heroic solitary explorers climb intellectual mountains, valiant discoverers race against one another to win a Nobel prize, intrepid teams celebrate advances and breakthroughs. Such glorified visions of the past are updated versions of classical myths. Stepping into the sandals of the gods, scientists have become the super-heroes of the modern age.[28]

When historians focus on famous individuals, they leave out many vital people who made science central to everyday life. Science's intellectual class system rates the achievements of gentlemen far higher than those of artisans and women. What about the technicians and administrators who made instruments work, recorded observations, collected and prepared specimens, catalogued results, organised the laboratory? The reputations of many celebrated men were built on the dangerous exploits and patient searches of countless explorers, collectors and classifiers. And surely tribute should be paid to those who provided financial support, or were knowledgeable enough to make constructive criticisms, check experimental readings and proofread manuscripts? Retrieving these invisible assistants – male and female – gives a far more realistic picture of how science was actually carried out.[29]

Another problem with heroic histories is that they isolate scientific achievements from the rest of society. As well as intellectual shifts, vast social transformations were also crucial for establishing the foundations of modern science. Scientific knowledge now dominates popular media as well as educational syllabuses, but this has only happened because teachers, editors, museum curators, translators and illustrators enabled other people to learn about new results and theories. Without them, science would have remained an esoteric scholarly pursuit, reserved for the privileged few. Many of these forgotten people were women.

Traditional histories of science focus on discoveries and inventions, almost invariably made by men who are elevated to the status of great heroes. Such models of progress may be appealing, but they distort the past by leaving out important parts of the story. Science cannot advance without a shared fund of knowledge. It is essential to make specialised publications accessible to wider audiences, both at home and abroad. Scientific progress depends on the existence of solid education systems, so that new generations of researchers can build on the results of their predecessors. Understanding how science has become so important means looking not only at its successes, but at how its achievements are perceived and disseminated. Because women were excluded from universities, scholarly societies and laboratories, they could not make the same contributions towards science as men. However, their work in education was vital for science's growth.

Science is not just a final product, such as a theorem, chemical or instrument, but is an integral component of society. Industry, business, warfare, government, medicine – they all depend on science and also affect how it develops. To understand how science came to form the backbone of our modern world, we need to describe not only what happened inside laboratories and studies, but also what happened outside.[30] Women as well as men have participated in the collective endeavour that brought about science's ubiquitous presence. Scientific history is not only about knowledge itself. It is also about how that knowledge was reached, taught and used. Broadening what counts as science's history entails recognising and crediting women's involvement.

Women made different contributions from men – but different need not mean insignificant. Women simply could not operate in the same way as men. They were educated separately, were seen as being physically and psychologically different, and were expected to assume domestic and educational responsibilities. In old-fashioned versions of science's history, great men stride along the road to truth, their achievements along the way marked by milestones of famous discoveries. Like lower-class men, women are inevitably excluded from these models of the past – rather like all those manly Victorian explorers who supposedly climbed to the tops of mountain peaks or reached remote river sources without the assistance of local guides and porters. Too many people have been air-brushed out from our visions of the past. Understanding how and why science has become so important means learning about artisans and assistants as well as about wealthy gentlemen, about science's integration into society as well as about esoteric theories and experiments – and about women as well as about men.

Without women, science would have developed very differently. For one thing, many important experiments would never have been started, let alone finished, without the unrecorded cooperation of wives, sisters and daughters. If women had not silently taken over the task, observations would have remained uncatalogued, collections would not have been classified, specimens would not have been illustrated. Without their female critics, many scientific books would be less polished, and some of them would not even have been written; others would have languished as piles of papers in a drawer, waiting for an editor to collate, organise and publish them. So many men must have presented collaborative endeavours as their own. As an English scientist wrote in a letter to his wife, 'Had our friend Mrs Somerville [a nineteenth-century physicist] been married to La Place or some other mathematician, we would never have heard of her work. She would have merged it with her husband's, and passed it off as his.'[31]

Just as significantly, women were vital for establishing effective channels of communication. To make science useful, ideas must travel in several directions, not just outwards from an elite core of specialists. Experimenters need to know which problems need solving, which lines of investigation meet public approval, which approaches would distinguish them from their competitors. Science is integral to modern society because a great deal of collective effort was involved in advertising its benefits and persuading governments and private investors to provide financial backing. Female educators often did far more than water down complicated ideas: they provided new interpretations, corrected mistakes, clarified obscure points and translated foreign books.

Without women, less would be known by far fewer people, and science would have eventually ground to a standstill. By translating important scientific texts into foreign languages, women enabled results to be rapidly transmitted between one country and another. They also engaged in an even more important type of translation – interpreting complex theories and translating them into straightforward explanations that could be understood not just by students, but also by the politicians and industrialists who direct the path of research through funding. Chauvinists still joke about women's (alleged) inability to tackle mathematics and science. In reality, women were crucial for enabling the spread of science, technology and medicine into many different areas. Without their participation, our lives today would have been very different.

If the past is a foreign country, then *Pandora's Breeches* is a new type of guide. Remapping the entire territory in a single trip – rewriting the whole of science's history in one book – would be an impossible project. Rather than setting off immediately into the unknown, *Pandora's Breeches* starts from the security of the familiar, but suggests different pathways to follow. Instead of completely abandoning the well-known landmarks of the past, those famous men and their heroic discoveries, this book examines some of them from other angles. By telling new stories about old characters, it reveals different ways in which women have affected the development of science.

One way of thinking about *Pandora's Breeches* is to consider Figure 6, an imaginary man-woman scientific couple that symbolises the real-life pairs presented in this book. Writing under a pseudonym, the caricature publisher Samuel Fores produced this image as part of his invective against male midwives, who were much in demand by fashionable mothers at the end of the eighteenth century. Fores wrote a venomous booklet designed to expose 'their cunning, indecent and cruel practices' and encourage women to withstand these upstarts taking over their traditional area of expertise. (No accident that a man should be somewhat unexpectedly defending the rights of female experts – Fores also provided helpful hints for husbands who were worried about the intentions of these male invaders entering their wives' boudoirs.)

First published in 1793, this print has been interpreted in several ways – much as *Pandora's Breeches* rewrites the relationship between real scientific partners from the past. Most obviously, Fores's picture compares two stereotypes. On one side stands the female midwife who points to the blazing fire heating up a pan of water. Her bright informal clothes and the flowered floor-covering send out comforting signals that birth is a natural event, when the mother will be eased through labour by experienced women clustered around her in the seclusion of her own room at home. The sharp dividing line sears down the page like a scalpel-cut. On its other side, the man's dark tailored jacket radiates authority and his aggressive instruments signal the pain ahead, when cold metal will grab the fragile baby and wrench it out.[32]

Men who ventured into women's traditional spheres were regarded with as much hostility as women who dared to engage in science. Over two centuries later, the medical management of childbirth is still being challenged. This bisected figure also symbolises angry conflicts between historians with different views about interpreting the past. For those with faith

Fig. 6
A Man-Mid-Wife.
John Blunt (Samuel Fores), frontispiece of his *Man-Midwifery Dissected* (1783),
hand-coloured etching.

The original hand-written caption reads:
'A man-mid-wife or a newly discover'd animal, not Known in Buffon's time; for a
more full description of the *Monster*, see, an ingenious book, lately published price
3/6 entitled, Man-Midwifery dissected, containing a Variety of well authenticated
cases elucidating this animals Propensities to crudity & indecency sold by the
publisher of this Print who has presented the Author with the Above for a
Frontispiece to his Book.'

in scientific progress, male midwives represent the triumph of clinical experts over gin-swilling, ignorant women like Charles Dickens's Sairey Gamp. Many feminists have disagreed. Far better, they insist, to encourage natural birth than artificial intervention, to rely on women's innate instincts than on icy expertise and metallic instruments. In addition, they continue, these male intruders elbowed women out of their rightful domain – men excluded women from the new medical specialities and forced them into subordinate positions below male professionals.

All these arguments portray women as hapless victims with little control over their own fate. Yet another way of interpreting this picture restores women to a position of power. Midwifery services had to be paid for, and wealthy women preferred to employ trained men with their instruments, instead of traditional female attendants. Male midwives became popular through female choice. Reinterpreted, Fores's dichotomous image proclaims the authority of women, not their subservience.[33]

There were many real-life pairs of scientific women and men – sisters and brothers, wives and husbands, patrons and protégés. Like reinterpreting Fores's imaginary duo, redrawing their linked relationships gives women a powerful presence. At the most basic level, many men would simply have been unable to cope without the emotional support of a companion who was sufficiently educated and intelligent to discuss setbacks. It is too easy to denigrate these countless women by dismissing them with empty clichés – handmaidens of science, or the power behind the throne.

Even towards the end of the nineteenth century, some scientists brought their research right into the home, traditionally a female realm. This meant that even women with no scientific training inevitably became involved. Day after day Charles Darwin's wife Emma nursed him through bouts of sickness, probably psychosomatic in origin, yet painfully disabling. She warded off unwelcome visitors, listened patiently through his moments of self-doubt, and inwardly sighed with relief when he found a new problem to consume his attention and hence calm his spirit. This fragile yet manipulative scientist converted Emma into a mother-figure, whimpering, 'Oh Mammy I do long to be with you & under your protection for then I feel safe.' In the evenings, they discussed his work together, and he recruited further women – other scientists' wives, Emma's friends – as his unofficial editorial team. Darwin's great book on evolution was a collective effort whose success depended not only on contributions from countless collectors and naturalists, but also on the female workers who made his ideas comprehensible.[34]

Darwin's science occupied the entire household. When one of their children visited a friend, he politely enquired where the barnacle dissecting room was – his father's unusual obsession was this boy's daily normality. There was no chance for Emma Darwin to retreat into private domesticity. In a typical project to investigate a carnivorous plant's feeding habits, Darwin took over one of the kitchen shelves as a laboratory and scoured the house for flies, spiders and bits of bathroom sponge. He used their children and pets as research subjects, sketching their expressions and making scientific notes on their rate of development; and he enrolled the young mothers in his extended family as research assistants to report on their own babies' behaviour.[35]

Despite this domination, Charles Darwin evidently saw himself as a liberated husband, comparing himself favourably with their friend Charles Lyell. Lyell, a lawyer who became famous as a geologist, effectively represented two people's work as one by relying on his wife Mary for scientific as well as emotional support. Even during their engagement, he coerced her into studying geology and learning German so that she would be useful to him later. He converted their honeymoon into a scientific field trip, and throughout their childless married life exploited her skills so that he could pursue a successful and productive career. She compensated for his deficiencies in several areas: she read to him for hours on end because of his poor eyesight, she translated foreign articles that he could not understand, she illustrated his books because he could not draw, she edited his writing to ensure it was stylishly written and error-free, she became more expert than him on conchology, and she classified his specimens to save him the trouble. She even recruited her maid: 'I have taught Antonia to kill snails and clean out the shells and she is very expert.'[36]

Information exists about women like Emma Darwin and Mary Lyell because their husbands were famous scientists and they were born into wealthy families. Many earlier women must have had similar experiences, although reconstructing their lives in any detail is much more difficult. The seventeenth and eighteenth centuries were crucial for the explosive growth of science during the Victorian era, yet less evidence of these concealed women survives.

It is hard to get away from heroic versions of history. For one thing, our vision of the past, our internal map, is structured by celebrated names and events – Galileo, Newton, Darwin, Einstein. Simply throwing away an old map is not the best way of assessing the value of a new one. Rather than setting out with a totally blank sheet, it is easier to navigate by keeping some familiar landmarks. In addition, writing new stories demands reassessing the evidence, and since historical records are generally clustered around prominent people, that means starting with those eminent men about whom we have more information.

So as a first step in rewriting scientific history, *Pandora's Breeches* concentrates on a small selection of women who were linked to famous men in different ways. These were unusual women, to be sure. But in traditional versions of science's past, even their attainments have been eclipsed by the auras radiating from men. Retelling the stories of selected man–woman partnerships does more than just restore these unusual women's lives to visibility. Breaking away from conventional history to give women more prominence reveals the activities, opinions and aspirations of other individuals about whom we have little specific information. For all the women who left behind tangible traces, there must have been many more who have disappeared from the archives. Evidence about women's lives is hard to retrieve, but their ghostly presence in the surviving records yields tantalising hints of their very substantial real-life existence.[37]

Chapter 2

Lady Philosophy/Francis Bacon

What thou seest in me is a body exhausted by the labours of the mind. I have found in Dame Nature not indeed an unkind, but a very coy Mistress: Watchful nights, anxious days, slender meals, and endless labours, must be the lot of all who pursue her, through her labyrinths and meanders.

Alexander Pope, *Memoirs of the Extraordinary Life,*
Works and Discoveries of Martinus Scriblerus, 1741

'Every picture tells a story.' This snappy slogan was coined in the early twentieth century to advertise Doan's Backache Kidney pills, but it could just as well be used to describe scientific pictures being drawn hundreds of years earlier. Many books were introduced by elaborate frontispieces; packed with coded signals, they summarised the author's central message. These images relied on a pictorial language of symbols and resonated with mythological, religious and contemporary allusions. Modern readers have lost the art of understanding these allegorical frontispieces, and so now they have to be translated into words. Although they were designed to be deciphered, they were also deliberately made ambiguous so that they could be read in different ways – and they still can. Historians can present different versions of the past by choosing how to interpret frontispieces and – just as importantly – deciding which ones are important to study.

Historians often reproduce the frontispiece shown here as Figure 7. It originally appeared in Francis Bacon's Latin book *Novum Organum* (*The New Organon*), which – along with some of Bacon's other books – became a major manifesto for scientific research throughout the seventeenth and

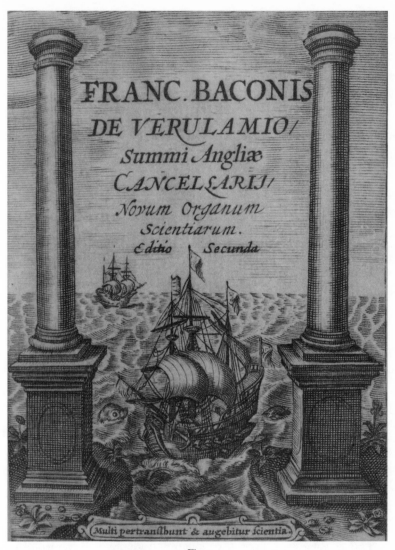

Fig. 7
Enlarging the empire of knowledge.
Frontispiece of Francis Bacon's *Novum Organum* (1620).

eighteenth centuries. *The New Organon* was first published in 1620, and Bacon chose his title to emphasise that he was rejecting the familiar scholarly logic of Aristotle's *Organon* and replacing it with his own experimental agenda. A lawyer by training, Bacon climbed to power during the reign of James I, eventually becoming Lord Chancellor. Although he did not carry out experiments himself, his programmes for progress provided the blueprints of modern science. As conventionally read, Bacon's frontispiece illustrates a traditional story about science's foundations, one that emphasises the significance of great men and their revolutionary innovations.

The mythological pillars of Hercules straddled the Straits of Gibraltar, gateway between the Atlantic and the Mediterranean. Christopher Columbus had demonstrated the rewards of venturing out from these well-charted waters, and now traders were bringing back plants, animals and other goods that stimulated the economy and provided new sources of information about the natural world. Similarly, Bacon wanted to send out ships of learning that would leave behind the safety of the Mediterranean classical world. 'Many will travel and knowledge will be increased,' reads the Latin quotation from the biblical Book of Daniel draped between the pillars' feet. Bacon prescribed an experimental agenda. Only through collecting and organising massive amounts of data, he taught, could the laws of nature be uncovered. This was to be a collaborative endeavour, one based on cooperation and communication. In order to achieve this research programme, Bacon envisaged a utopian community based on the island of Atlantis, where Salomon's House, a new temple of wisdom, would be dedicated to investigating ways of harnessing nature's powers for the benefit of society.

Hercules's pillars were inscribed with a warning for adventurous sailors: *ne plus ultra* – sail no further, or nothing lies beyond. Bacon's frontispiece could well be summarised as *plus ultra*, the defiant opposite of this traditional slogan. And this was the title coined by Joseph Glanvill, a clergyman from Bath, for a book emphasising his Baconian allegiance. Glanvill was one of the earliest propagandists for London's Royal Society, which was established in 1660 as a flagship of Baconian ideology. Keen to advertise their novel approach to learning, the founding Fellows believed that the best 'way to support our owne Enterprise is to devise all wayes to revive Lord Bacons lustre'.[1]

After only seven years, the fledgling Society's first history appeared. Since there had not yet been much time for great progress, its author – Thomas Sprat – made grandiose claims for the achievements that their Baconian enterprise would make in the future. For his *History of the Royal-Society*, Sprat picked the frontispiece shown in Figure 8, which is now one

Fig. 8
Baconian ideology
at the Royal Society.
Frontispiece of Thomas
Sprat's *History of the Royal-
Society* (London, 1667).

of the best-known images from early modern science. On the left, the
Society's first President points towards King Charles II, who is being
crowned with a wreath of laurel by the goddess of fame. Diplomatically,
Charles has been given the most prominent position in the hope of securing
further royal patronage, preferably in the form of financial support.
Nevertheless, the Society's ideological figurehead is Bacon, sitting on the
right in his Chancellor's robes and pointing to the instruments that from
now on should be the true source of knowledge. Through experiment and
free discussion, the Fellows' inventions would, they claimed, improve
human welfare throughout the world. The building glimpsed through the
archway on the right could well be a design for Bacon's idealised research
centre, the House of Salomon. On the left, the shelves overflow with books
by the latest scientific authors – William Harvey, Copernicus, Bacon himself
– and in the background lie two of the Royal Society's crowning achievements, a giant telescope and the air-pump designed by the aristocratic
chemist, Robert Boyle.

There is nothing untrue in this version of how these leisured gentlemen designed big instruments and adopted Bacon as their figure-head (although historians have concluded that this famous frontispiece was designed for a different project and was not even included in every copy of Sprat's *History of the Royal-Society*[2]). But because the past and its pictures are open to interpretation, different stories can be told about the early Royal Society, ones that stress continuity rather than abrupt change, and that take the spotlight away from heroic men and their splendid discoveries.

One of these other versions of the past emphasises the importance of the concealed assistants and artisans who made the new instruments and experiments function properly. Although their existence is virtually invisible in published reports, the gentlemanly Fellows of the Royal Society relied on their hard work. In Figure 8, two prestigious new inventions – Boyle's air-pump and the telescope standing behind it – are prominently placed next to the King's head, yet they could never have been made without the manual labour of back-room technicians who have been obliterated from conventional heroic histories.

Yet another way of thinking about science's history is to pay more attention to everyday activities. Because historians have focused on large and momentous discoveries, they have ignored the numerous ordinary instruments that are displayed on the walls; the ones sitting on top of the bookcase might even suggest that they were more important than the volumes on the shelves. Most of this apparatus was made by modifying traditional measuring, surveying and navigation devices – the tools that expert craftsmen had developed over the centuries, and which were far more vital for experimental research than the esoteric theories discussed by learned philosophers. Skilled artisans had built up the practical expertise lying at the heart of the new science.[3]

The early Royal Society claimed to be democratic, an open, neutral society that made no discrimination on grounds of religion, nationality or occupation. In reality, it was an elitist organisation dominated by educated aristocrats and landowners who formed a new scientific priesthood. According to Sprat, the Fellows wanted to establish an active research centre that would receive contributions 'not onely by the hands of Learned and profess'd Philosophers; but from the Shops of *Mechanicks*; from the Voyages of *Merchants*; from the Ploughs of *Husbandmen*; from the Sports, the Fishponds, the Parks, the Gardens of *Gentlemen*'.[4] But these data-gatherers

– ants, as Bacon called them – lay at the bottom of an intellectual hierarchy. At the top lay natural philosophers – Bacon's wise bees – who used their minds to digest experimental results. Information was solicited from many sources, but the task of sifting through and collating it was reserved for privileged gentlemen.

Despite the propaganda, in practice the Royal Society was a closed institution, a formalised version of gentlemen's coffee clubs with limited access. The King never responded to all those hints about money, and so – unlike in France – scientific research was not funded by the state. To provide financial stability, an annual fee was imposed, which immediately deterred all but the most wealthy. Furthermore, because cross-country transport was slow and expensive, provincial Fellows could only participate through correspondence. And although not *explicitly* excluded, half of the country's population was ineligible for membership – Sprat evidently found it so unthinkable that women might want to join the Society that he forgot to ban them.

The Fellows were forced to confront this possibility in 1667, when Margaret Cavendish (see Figure 4) announced that she wanted to visit the Society and watch some experiments being performed. Because she was married to the powerful Duke of Newcastle, and was known to be a generous patron of Cambridge University, it was hard to refuse her request. On the other hand, not only was she a woman, but she had also written books criticising the experimental programme of research being promoted by Boyle and his colleagues. What use is it, she demanded, to admire bees through a microscope if this knowledge is not used to produce more honey? Cavendish condemned the 'strange Conceits' of men, accusing them of reducing vital nature to a lifeless object of study, 'a dull, inanimate, senseless and irrational body'. Some Fellows may have agreed with her protests, but Cavendish had acquired a reputation for eccentricity, and many of the men (and their wives) were more interested in her good looks and fine clothes than in her controversial ideas. In the end, Cavendish was allowed to attend one single session when Boyle – presumably obliged to conceal his disdain – performed some spectacular experiments for her entertainment.[5]

As Samuel Pepys noted in his diary, arguments about her at the Royal Society were heated, and Cavendish remains the subject of debate.

Previously dismissed as 'Mad Madge', during the last few decades she has acquired a new status as a suppressed proto-scientist. Feminists have converted this former figure of fun into a martyr, the hapless victim of male oppression. Yet this solution also seems unsatisfactory. The sacrificial heroine may be an appealing image, but it is too simplistic. Cavendish's books reveal that she was no intellectual heavyweight, and her whimsical style and incoherent arguments must have prejudiced even sympathetic Fellows against her. Better-educated women sneered at Cavendish's affectations, and worried that her ostentatious activities would reinforce prejudice against genuine female scholars. Nevertheless, this notorious episode does symbolise the exclusion of women from science. Following Cavendish's invasion with her retinue of elegant ladies, no women were elected to full membership of the Royal Society until 1945.

So the problem remains: how can science's early history be told in a way that realistically takes account of women's involvement? Figure 9 shows a frontispiece that is rarely reproduced, but suggests a different approach towards interpreting Baconian projects of research and improvement. This allegorical vision of natural philosophy was designed for the English translation of an Italian book that had been published in 1667, the same year as Sprat's *History of the Royal-Society*. Produced by the Academy of Cimento in Florence, it contained detailed reports on a topic that also preoccupied the London Society – examining how things behave in a vacuum. This avenue of research had only recently been made possible by the invention of the air-pump, and it was very controversial. It involved creating an artificial state of nature in order to investigate normality, what Bacon had called 'twisting the lion's tail'.

The societies in Florence and London had very similar aims, but this frontispiece is – in marked contrast with Sprat's – scarcely seen in English history books. On the right sits London's Royal Society, ready to receive the volume of experimental results being respectfully tendered by the Florentine Academy. The man on the left of the picture is an elderly Aristotle. Hardly surprising that he looks offended – he is being informed by a nearly naked Nature that his teachings have been overthrown by the Baconian experimentalism of these two youthful societies. As she pulls back her cloak, Nature implies that her secrets will be uncovered by

Fig. 9
Knowledge is power.
Frontispiece of *Essays of Natural Experiments, Made in the Academy del Cimento,* translated by Richard Waller (London, 1684).

London's natural philosophers, who have now inherited the Italian discoveries. She seems to be illustrating Sprat's own delight that 'the Beautiful Bosom of *Nature* will be Expos'd to our view'. Experimental philosophers, he preached, should not hold back in their 'Courtship to Nature' because – like any mistress – she 'soonest yields to the *forward*, and the *Bold*'.[6]

Women were officially excluded from participating in experimental research, yet here both societies are represented by glamorous damsels draped in classical robes. The rules for drawing women to portray the different spheres of knowledge had been laid down at the end of the sixteenth century, and 200 years later art teachers were still publishing revised versions that carefully explained which features should characterise each academic discipline or personal quality. Artists could turn to their handbooks to find the appropriate female figures for all the sciences studied by young gentlemen, such as astronomy, geometry and geography. Women carrying suitable attributes also symbolised subjects that have now disappeared from the educational syllabus – rhetoric, logic, eloquence. Even manhood was depicted as a young woman sitting on a lion.[7]

The queen of all these sciences was Lady Philosophy, conventionally portrayed – as in Figure 10 – sitting on a marble throne, holding open the books of natural and moral knowledge. She encapsulated an important message for young gentlemen – careful study is the way 'to demonstrate the secrets of nature'. When he was grappling with the laws of gravity, Newton complained: 'Philosophy is such an impertinently litigious Lady that a man had as good be engaged in Law suits as have to do with her.' At the time, his friend Edmond Halley (now famous for his comet) was trying to coax Newton into writing down his new theories for his major book on gravity. Worried that Newton might stagger to a complete

Fig. 10
Lady Philosophy.
Figure 120 of George
Richardson's *Iconology*
(London, 1779).

halt, Halley soothingly dissuaded him from 'desisting in your pretensions to a Lady, whose favours you have so much reason to boast of'. Lady Philosophy taunting male researchers: the apparently jocular interchange of these two men reflects centuries of symbolism.[8]

But although Lady Philosophy was queen of the allegorical realm, in the kingdoms of western Europe it was men who carried out the experiments to decipher her book of natural secrets. This divorce between practice and imagery was emphasised by Abraham Cowley, a well-known literary writer and one-time Royalist spy, in a poem that formed the preface to Sprat's *History of the Royal-Society*:

> Philosophy, I say, and call it, He
> For whatso'ere the Painters Fancy be,
> It a Male Virtu seems to me.[9]

A close friend of Sprat's, Cowley articulated poetically the sentiments felt by many Fellows. Placed at the very beginning of Sprat's propaganda piece for the new Royal Society, this verse reinforced the accepted notion that all the practitioners of the new natural philosophy were to be men. But it also made a much stronger statement, arguing that this division between the sexes prevailed in natural philosophy at a deeper level, right at the heart of scientific discovery. Never mind, Cowley maintained, that Philosophy is artistically shown as a woman. The knowledge that these men would gain was itself gendered masculine – 'a Male Virtu', as he put it.

English Baconianism suited men who aimed to govern. 'For knowledge itself is power,' Bacon had declared, a memorable slogan that was often repeated during the following centuries. For the scientific programme that he launched, knowledge meant not only power over nature, but also power over people – including aristocrats exploiting their workers, England ruling her colonies, and men dominating women. As Boyle explained, 'there are two very distinct ends, that men may propound to themselves in studying natural philosophy. For some men care only to know nature, others desire to command her . . . and bring nature to be serviceable to their particular ends, whether of health, or riches, or sensual delight.'[10]

The new experimental philosophy at the Royal Society was to be an agent of civilisation that would help England to rule over an intellectual empire. And it was also a specifically male enterprise, one that would perpetuate the dominion men enjoyed over women and create an intrinsically masculine form of science. Sprat sneered at 'the *Feminine* Arts of *Pleasure*, and *Gallantry*' to be found in Europe, and rejoiced that 'the *English Tongue* may also in time be more enlarg'd, by being the Instrument of conveying to the World, the *Masculine* Arts of *Knowledge*'.[11]

Misogyny was, of course, nothing new in western Europe. For people of the seventeenth century, whose major source of learning was the Bible, women's subordinate role had been ordained right from the creation of the world. Only as an afterthought had God made Eve from one of Adam's ribs – she was a second-best, an accessory designed to meet Adam's needs. Further removed than him from God, she was also more prone to sin, and more susceptible to the serpent's temptation. Because of Eve's weakness, Adam bit into the fruit from the tree of knowledge, a defiance of God's orders that led to the couple's expulsion from the Garden of Eden. For centuries, this story of the Fall had been used to justify women's subjection. Men's scientific insight was clouded by their emotions, lamented one Baconian apologist, because 'the woman in us, still prosecutes a deceit, like that begun in the Garden: and our understandings are wedded to an Eve, as fatal as the mother of our miseries'.[12]

Women were traditionally equated with nature. Figure 11 displays one of the most common stereotypes – the female nurturer. Entitled *Man Born to Toil*, this picture is the first in a series of six allegorical plates extolling the rewards of labour and diligence, and was produced when Bacon was a boy by a prominent Dutch artist, Maarten van Heemskerck. In case anyone misses the point, van Heemskerck has placed the label 'Nature' by the feet of the scantily draped woman. Her multiple breasts will nourish the infant she is holding so protectively; the serenity and curved lines of her pose mirror the tranquil, pastoral scene of cows grazing in lush, gently rolling fields. Such idealised visions of maternal nature reinforced declarations that women belonged at home, raising children and maintaining domestic harmony, just as nature was benevolent, nurturing and at the service of mankind.

Nevertheless, the picture's backdrop is more threatening. The bare mountain peaks, the hovering predatory bird and the stormy, swirling clouds all suggest that nature can be cruel, unpredictable and dangerous.

Ales et à primis producit in ora nidis
Iam iam plumantes certo modulamine fœtus,
Hortaturque sequi, breuibusque insurgere pennis;

Sic genus humanum rerum Natura nougtrix
Mollibus è cunis, grauidae parentis ab aluo,
Ducit ad ærumnas, et duros cauta labores.

Fig. 11
Nature and the mechanical world.
Maarten van Heemskerck, *Man Born to Toil* (1572), engraved by Philips Galle.

Like a flickering hologram, this alternative stereotype flourished simultaneously with the mother-earth figure. Identifying women with the disor-
derly, unruly aspects of nature underpinned images of women as voracious
sexual predators or diabolic witches. Women needed to be controlled and
subdued in order to prevent their primitive urges from disrupting social
order.

These complementary versions of female nature provide a stark contrast
to the globe covered with mechanical devices and craftsmen's tools. As the
Latin verses beneath the picture imply, mother nature will force her fledg-
lings to leave the soft maternal nest and learn to survive in the hard life
of work.[13] To deliver this Protestant message about the rewards of industry,
van Heemskerck has obviously studied the implements being sold by Dutch
craftsmen, then amongst Europe's finest. Angular instruments entrap the

43

curved Earth, constraining her into a perfectly spherical shape. Some of them – the hoe, axe and hammer – are conventional images for manual labour. But others belong to the new practice of natural philosophy. One type – such as the compass-dial, set squares and hour glass – measure and record nature's vicissitudes, and help to construct regular laws describing her behaviour. Yet others – the butterfly net, the scissors, the saw – reveal her behaviour only by trapping her, breaking her down into pieces, stripping away protective surface layers.

Bacon may never have seen this particular picture, but it does capture the ambiguous feelings towards women and nature that prevailed in Europe when he was formulating his plans for progress. Like other influential writers, Bacon became an ideological figurehead because he articulated opinions that were held by his contemporaries. He wrote the right books at the right time, and his successors could form interpretations of his work that suited their own agendas. But Bacon also brought his own ambitions and motivations to his project of reform, and his political and legal career ensured that his individual mark permeated the legacy he bequeathed to natural philosophy. By contributing his own powerful imagery, derived from his personal experiences, Bacon enabled gender prejudices to be knitted into the very fabric of science.

In common with many of his colleagues, Bacon was preoccupied with the implications of the Fall in the Garden of Eden, and saw science as a means for humanity to redeem itself. Instead of trying to fathom the mysterious mind of God, natural philosophers would unearth the secrets of nature – like new Adams, he promised, they would rectify the damage wrought by Eve. Glanvill preached that men should purify themselves of any inner corrupting female influences. 'The *Woman* in us,' he worried, 'still prosecutes a deceit, like that begun in the *Garden*; and our *Understandings* are wedded to an *Eve*, as fatal as the *Mother* of our *miseries*.' Controlling nature also entailed controlling oneself. Women, so the ideology ran, were governed by their passions rather than by their reason. If men were to succeed as natural philosophers, then they should suppress the feminine, emotional aspects of their character so that their intellectual powers could flourish. Judging things by 'the *Gusto* of the fond *Feminine*' would, Glanvill warned, never enable men to approach 'the *Tree of Knowledge*'. Scientific research would be impossible if 'the *Affections* wear the breeches and the *Female* rules'.[14]

To prescribe how mastery over nature should be achieved, Bacon coined imagery from the witchcraft trials in which he was involved. He organised his own bid for power under James I, the Scottish king responsible for England's vindictive campaign of persecution against women who were denounced as witches for challenging authority. In his works on natural philosophy, Bacon described how nature's secrets should be wrested out of her just as confessions of witchcraft could be extorted from women by tools of torture. National alarm about female insubordination was fuelled by rumours that witches fornicated with the devil, and sexual imagery pervaded Bacon's discussions of scientific investigations. Addressing the King, Bacon drew a vivid parallel between the procedures of extracting truth from nature and from women: 'For you have but to follow and as it were hound nature in her wanderings . . . Neither ought a man to make scruple of entering and penetrating into these holes and corners, when the inquisition of truth is his whole object, — as your Majesty has shown in your own example.'[15]

Like all metaphorical writing, Bacon's language incorporated complex, ambiguous images. In the unpublished guide that he composed for 'my son', an imaginary successor, Bacon explained how progress could be achieved: 'I am come in very truth leading to you Nature with all her children to bind her to your service and make her your slave.'[16] In one extraordinary vision in this manuscript, Bacon dreamed of a future in which the union of celibate natural philosophers with nature herself would ensure that his ambitions were realised: 'My dear, dear boy, what I purpose is to unite you with *things themselves* in a chaste, holy, and legal wedlock; and from this association you will secure an increase beyond all the hopes and prayers of ordinary marriages, to wit, a blessed race of Heroes or Supermen who will overcome the immeasurable helplessness and poverty of the human race.'[17]

Bacon's inheritors gleaned his work selectively, oblivious to some of his references, yet picking out and developing ideas that suited their own interests. Bacon did not invariably depict female nature being overpowered by male investigators, but sexually charged language of mastery did get woven into scientific propaganda. Newton and his fellow students learned from his Cambridge mathematics professor that the aim of natural philosophy was to 'search Nature out of her Concealments, and unfold her dark Mysteries'. When Newton himself was famous, the astronomer Edmond Halley congratulated him on 'penetrating so far into the abstrusest secrets of Nature'.[18]

Imagery of penetration, undressing and domination permeated advertisements for the early Royal Society. Experimenters portrayed themselves as nature's philosophical suitors, who achieved success when 'you woo your mistress with boldness and importunity . . . the surest and most powerful way to win her . . . many others proceed with too much hesitation and caution.' Contemplation alone, they warned, would never persuade 'Dame Nature' to 'unlock her Cabinet'. Instead, those who wanted to 'penetrate into Nature's antechamber to her inner closet' should use manipulative instruments like Boyle's air-pump, which distorted nature into an artificial vacuum. This Baconian invention would, its advocates promised, allow an experimental 'inquisition' of nature to wrench out 'a confession of all that lay in her most intimate recesses'.[19]

Metaphors evoke unsuspected connections, refuse to let meanings be pinned down. They add richness, but make language fluid so that new generations of readers bring new interpretations. A hundred years after the Royal Society's foundation, the doctor Erasmus Darwin (grandfather of Charles) described his recent enthusiasm for hunting fossils in a letter to his close friend, the potter Josiah Wedgwood, who had a professional interest in seeking out new types of minerals. Darwin held a traditional image of the Earth as a woman's body. Geological excitement seems to have overwhelmed him when he reported that 'I have lately travel'd two days journey into the bowels of the earth and have seen the Goddess of Minerals naked, as she lay in her inmost bowers, and have made such drawings and measurements of her Divinity-ship, as would much *amuse*, I had like to have said *inform*, you.'[20]

Darwin would not, of course, have written such a comment in a serious scientific article. His sense of humour now seems rather strange, but trying to appreciate the witticisms of a bygone age is a marvellous route to understanding alien patterns of thought. Metaphors may be only figures of speech, but they do affect how the world is perceived. After the first person said that 'the lion is the king of beasts', people thought differently about kings as well as about lions. Similarly, conceptualising nature as a beautiful woman who needs to be undressed inevitably affected how scientific experimenters viewed their work.

Bacon had wanted to seize nature and make her 'condescend to unveil for us her mysteries'. The veil of nature was another potent metaphor elaborated by Bacon's successors. Made of semi-translucent material, veils both conceal and reveal. Shimmering allusively between multiple references, veils were a favourite image of Gothic novelists, and were also used to great effect in the public lectures of the chemist and poet Humphry Davy in the early nineteenth century, almost 200 years after Bacon's death. Davy promised huge future rewards from scientific investigation since, he argued, nature's disrobing had scarcely started: 'The skirt only of the veil which conceals these mysterious and sublime processes has been lifted up, and the grand view is as yet unknown.' As the Victorian era closed, the same language prevailed. One of Britain's leading scientists, Sir William Crookes, urged researchers 'to pierce the inmost heart of Nature . . . Veil after veil we have lifted, and her face grows more beautiful, august, and wonderful.'[21]

Virtuous women would wish to hide themselves away from the gaze of men, yet the language of veils suggests that scientists should strip nature's protective clothing away against her wishes. But women had also been given a reputation for flirtatiousness: the female Nature in Figure 9 is deliberately, even arrogantly, pulling aside her cloak and displaying herself to Aristotle. In addition, she is exposing her body to the book's readers, seventeenth-century natural philosophers – almost exclusively men. Such images (and there are many of them) are deeply ambivalent and suffused with sexual frissons. Does the veil imply that nature is chaste, like a nun or a bride? Maybe the tantalising veil conceals a virago, or a sight too ugly to be revealed. Or is Dame Nature an Eve-like figure, tempting men to probe into mysteries they are destined never to uncover?[22]

This quandary has been central to Western science since its foundations, and is no nearer resolution. Nature is still female, still mysterious, still a dangerous witch-like opponent to be explored, penetrated and conquered. As a modern Nobel prize winner explained, 'the experienced scientist knows that nature yields her secrets with great reluctance and only to proper suitors'. Female nature refuses to submit, repeatedly breaking the laws that rein her in by producing unpredictable events which force scientists to admit their limitations. In 1980, one geologist conceded defeat after a volcano unexpectedly erupted: 'Her flanks are shuddering . . . We don't know her intentions. Scientists haven't been able to probe her deeply enough with their instruments.'[23]

Many women – and men – object to this imagery lying at the heart of science, and feminist philosophers have challenged the gendered nature of scientific knowledge.[24] Karl Marx claimed the point of philosophy was to change the world, not just interpret it, and there are many ways of trying to do so. *Pandora's Breeches* is primarily a book about history, not philosophy, but reinterpreting the past can also help to improve the future. Understanding how prejudices have been built into science is important because it can prompt people to alter their current behaviour. Even the most egalitarian scientists need to learn not only about how women have been prevented from participating in science, but also how chauvinist attitudes pervade research at still deeper levels.

The events of the past cannot be changed, but new versions of what happened can be written. New facts can be discovered, and ones that were previously known can be given a new significance. There is no single definitive history of science – or, indeed, of anything else. Science is central to today's society, but appreciating its rise to eminence involves far more than just celebrating its instruments and its theories. The story of science is too often told as an exciting race between brilliant men who pledged themselves to the cause of truth. Such romanticised tales of heroic struggle imply that science succeeded simply because it was right. They ignore how science was disseminated, advertised and applied; they eliminate the practical expertise built up over centuries, which was vital for experimental investigations; and they obliterate the many, many thousands of men and women who have contributed to science's growth over the last few hundred years. Science developed because of collaborative efforts, and rewriting science's history must also be a collective project. *Pandora's Breeches* is one contribution.

In the Shadows of Giants

But censure is perhaps inevitable; for some are so ignorant, that they grow sullen and silent, and are chilled with horror at the sight of any thing, that bears the semblance of learning, in whatever shape it may appear; and should the spectre *appear in the shape of* woman, *the pangs, which they suffer, are truly dismal.*

Elizabeth Fulhame, *An Essay on Combustion, with a View to a New Art of Dying and Painting*, 1794

'If I have seen further it is by standing on ye shoulders of giants,' declared Isaac Newton. Boasting through false modesty, and keen to be acknowledged as the world's leading natural philosopher, Newton borrowed a familiar biblical image. Did he ever dream that this neat disclaimer would become one of science's most famous quotations, a favourite slogan for scientists celebrating their triumphant march up the mountain of truth? Newton is now, of course, regarded as one of the three great giants of his own age, partnered by his French predecessor René Descartes and his German contemporary, Gottfried Leibniz.[1]

Figure 12 shows the frontispiece of a physics text that tackled a hard task – explaining the ideas of Newton and Leibniz to French students. Published to great acclaim in 1740, this scholarly book was written by Émilie du Châtelet, a woman whose life has been overshadowed by the greatest giant of Enlightenment history, her lover Voltaire. In the centre, climbing up towards the Temple of Reason, is a solitary female figure, probably du Châtelet herself. Echoing Newton, she wrote: 'We bring

Fig. 12
The Temple of Reason.
Frontispiece of Émilie du Châtelet, *Institutions de Physique* (Paris, 1740).

ourselves up to the knowledge of truth like those Giants who scaled the Heavens by climbing on one another's shoulders.'[2]

At the bottom of the picture sits winged Astronomy, resting her arm on a celestial model and holding her measuring dividers in the other hand. She is accompanied by Agriculture watering a plant, and by other scientific muses gazing up at the radiant, naked figure of Truth, whose powerful beams of light pierce through the dark clouds of ignorance. These women are allegorical figures who only symbolise the power of knowledge. Just as Minerva represented wisdom, other classical goddesses personified nature and the sciences. Astronomy, chemistry, music, mathematics: all the disciplines were traditionally represented by female figures carrying the tools of their trade – dividers, balances, harps, and so on.

Du Châtelet's muses inhabit a fantasy world that is glimpsed through a solid rococo frame of reality. Here the ornate surround is surmounted by the portraits of three men – Descartes, Leibniz and Newton. In contrast with their clearly defined features, the ambiguous woman approaching the Temple of Reason does not reveal du Châtelet's identity, and her name does not appear on the title page. Female goddesses stood for scientific rationality, but in the real world, it was men who achieved it. Descartes, Leibniz, Newton: a powerful trio of men who look down from their superior position and who also loom over science's history, casting long shadows which conceal their contemporaries. Posterity has elevated them to fame, yet ignored the achievements of many, many others. In particular, the reputation of each of these three heroes was boosted by a woman.

Early modern Europe lacked telephones, aeroplanes and computers, but nevertheless there was an efficient communication network linking scholars and societies together. People wrote to each other, visited each other's homes, and discussed the latest theories, pictures and books. Descartes, Leibniz and Newton lived at different times (but overlapping) in different countries, yet they had many colleagues in common, either through letters or face-to-face meetings. They belonged to an intellectual elite which formed only a narrow band of society, yet stretched internationally. Many of these friendly circles were either exclusively male or exclusively female, but they did of course interact. Descartes, Leibniz and Newton all knew women, even though many biographies are written as though they lived in a world composed entirely of one sex – men.

The foundations of modern science were not built by male geniuses labouring in isolation. One way of thinking about it is to imagine each of these three men lying at the centre of a circle of contacts. Like intersecting ripples on the surface of a pond, their lives interlocked with each other because they had many acquaintances in common. But who is important depends on whose viewpoint you adopt – and each and every one of these acquaintances regarded themselves as sitting at the hub of their own circle. Historians choose which centre to prioritise, which network of contacts to emphasise, which interconnections are too remote to be significant. By concentrating on different links, they can present a different version of the past.

Chapter 3

Elisabeth of Bohemia /René Descartes

TO THE MOST SERENE PRINCESS ELISABETH, ELDEST DAUGHTER OF FREDERICK,
KING OF BOHEMIA, COUNT PALATINE, AND ELECTOR OF THE HOLY ROMAN EMPIRE
. . . And that this zeal is indeed in your Highness is obvious from the fact
that neither the distractions of the court nor the customary upbringing which
usually condemns girls to ignorance could prevent you from discovering all the
liberal arts and all the sciences . . . And I personally have a greater proof of
this, since I have so far found that only you understand perfectly all the trea-
tises which I have published up to this time. For to most others, even to the
most gifted and learned, my works seem very obscure.

René Descartes, *Principles of Philosophy*, 1644

Descartes, reported his niece Catherine, was a close friend of Lady Philosophy. In the picture reproduced as Figure 13, Lady Philosophy, identifiable by her crown of stars (representing physics) and the coiled snake biting its own tail (metaphysics), leads Descartes by the hand towards the radiant goddess of Truth. In front of Truth, a helmeted Minerva aggressively fights down the monsters of Ignorance and Prejudice. Scientific instruments are piled up on the ground, and eminent Greek sages cluster behind Descartes, who can be recognised by his face, as well as from the sheet he is holding, which describes his physical theories – whirlpools of tiny particles that swarm through every nook and cranny of the universe.

This engraving has had a chequered history. It was originally designed in 1707 as the frontispiece for a doctoral dissertation, but a senior academic

Fig. 13
Truth sought after by the philosophers.
Bernard Picart, *La Vérité recherchée par les philosophes*, 1707 (designed as the
frontispiece for *Thèse de philosophie soutenue par M. Brillon de Jouly, le 25 juillet, 1707*).

wrongly informed the Archbishop of Paris that it showed Aristotle wearing ass's ears and being vanquished by the new Cartesian philosophy. Such flippancy was sacrilegious for conservative Catholics. They associated Descartes's ideas with Protestant heresies, and suppressed the print. Some years later, it was published by an English plagiarist who removed all traces of Cartesian whirlpools and simply substituted Newton's name for Descartes's – scientific heroes are evidently interchangeable![1]

This particular Lady Philosophy had a double identity. She was also Christina, Queen of Sweden, renowned throughout Europe for her learning. By her early twenties, Christina was already being celebrated as the Minerva of the North, and engravings circulated showing her in a laurel-wreathed helmet with an owl at her side. As part of her ambitious plans to reign over an intellectual European empire, she encouraged eminent scholars to join her circle. Like other monarchs, Christina used her wealth and position to affect the directions of academic research, and during her reign she tried to convert remote Stockholm into a metropolitan centre fit to rival Paris and London. In 1649, she invited Descartes to visit her court. Five months after he arrived, he was dead.[2]

Descartes died in Sweden in 1650: a neutral statement of facts about a death shrouded in rumours. Did Descartes succumb to cold and pneumonia, or was he poisoned to prevent him from converting the Lutheran queen to Catholicism? Perhaps, as Catherine Descartes suggested, Dame Nature was so angry at being stripped naked that she unleashed a torrent of poison to push him into the grave. And what has happened to his saintly relic, the right forefinger lovingly chopped off by the French Ambassador? And whose skull was added to his remains as they were shipped around Europe?[3] Even his death has generated mysteries, so it is not surprising that his life has acquired a mythical aura. The traditional biographies resemble a philosophical fable . . .

Three women were important for René Descartes: his mother, who gave him life; Lady Philosophy, the bride to whom he pledged himself; and Christina of Sweden, who took his life away again. (Kitschy though this sentence deliberately sounds, it is a relatively prosaic version. Throughout the nineteenth century, Descartes's biographers described a fourth woman – a mechanical life-sized doll, eerily resembling his illegitimate daughter, who accompanied him all over Europe as tangible evidence of his edict that animals are machines without souls.)[4]

During his twenties, Descartes travelled widely, although there are strange gaps when his whereabouts remain unknown. He was only twenty-five when a series of bizarre, half-waking dreams in an overheated room gave him his life's mission: to rebuild the world of knowledge. Just as he had learned to conquer his nightmarish visions by the power of his own mind, so too, he claimed, rational thought would vanquish super-stition and solve the mysteries of nature. Like demolishing an ancient rambling city, Descartes set out to undermine the foundations of learning and construct an entirely new system based on reason. In his revolu-tionary rational philosophy, ideas about the world would be underpinned by facts as certain as mathematical truths.

Descartes coined philosophy's snappiest axiom: *I think, therefore I am.* I know that I exist, he argued, because I know that I think. Descartes split the universe into two. On one side lay matter – not only the phys-ical world of rocks, skies and oceans, but also the human body. On the other lay God, the angels and human minds. This Cartesian mind-body separation leads to a tricky problem: how can our brains interact with our bodies? Even Descartes admitted to feeling pangs of hunger when he needed physical food to nourish his intellectual activities, and – despite introducing God and the pineal gland into his system – he never did explain how our thoughts can govern our actions, or conversely how our sensations can influence our decisions.

Nevertheless, Descartes has often been hailed as the father of modern philosophy. His distinction between the human spirit and material objects is fundamental to one type of scientific ideology: if an observer is distinct from his surroundings, then he can observe the external world objec-tively. This Cartesian detachment, which implies a superior vantage point, is now often said to be masculine, because Descartes's harsh rational and mathematical approach encapsulates many of the characteristics that are attributed to men, and rejects qualities that are traditionally seen as femi-nine. His system overturned older organic cosmologies that regarded living beings as integral components of a harmonious universe, which bound people's minds and souls tightly together with God as well as with the physical world. Another way of thinking about this shift is to look at art. Medieval pictures, with their multiple scenes that depict inner as well as outer experiences, seem distorted and incoherent to us. Renaissance artists introduced a strong perspectival structure, which assumes that a single spectator is looking from a fixed external viewpoint. These pictures are reassuringly familiar and realistic because we have become used to representing the world in a Cartesian way.[5]

But in the mid-seventeenth century, Descartes's ideas were enormously controversial. For his English critics, Descartes epitomised the men of reason condemned by Bacon - not the wise bees, but the dogmatic spiders who spin webs from their own selves. His Catholicism posed further difficulties. Like Galileo, who was condemned by the Inquisition in 1634, Descartes taught that the sun, rather than the Earth, lay at the centre of our planetary system. Moreover, his concepts of matter contradicted the doctrine of transubstantiation (that consecrated bread and wine become the body and blood of Christ). Threatened with persecution, Descartes resumed his travels round Europe, moving from city to city as he became involved in a series of acrimonious disputes.

He eventually retreated to the Netherlands, taking with him only the scantiest of luggage – a Bible and a book by Aquinas. Isolated from the busy Parisian world, and lacking the financial and social support of a powerful patron, Descartes dedicated himself to a life of reflection. He was in his mid-forties when he published his philosophical masterpiece, the *Meditations*, an eloquent exposition based on those disturbing dreams that had launched him on his personal journey of discovery from doubt to certainty.

Buffeted by theological disputes and academic intrigues, Descartes found it impossible to turn down Queen Christina's invitation to Stockholm. Using the French Ambassador as intermediary, she had been writing to him for more than two years, and he was obliged to accept this summons from an influential, wealthy patron who demanded to be taught the most modern philosophy. Brought up as a boy, Christina resembled an intellectual cavalryman. Indifferent to clothes, food and discomfort, she devoted herself to her favourite pursuits – horse-riding, politics and academic study. Surviving on only five hours' sleep a night, Christina spurned female company and spent lavish sums to attract Europe's finest scholars to her court.

Demanding similar dedication from her new tutor, Christina established a rigorous timetable: five hours of lessons three times a week, to start at five in the morning – the depth of night in winter so far to the north. This was a hard regime for the man who gained his greatest insights dozing in a hot room, especially when the winter turned out to be the worst for sixty years. Descartes loyally struggled on for a couple of weeks before falling ill. Suspiciously refusing treatment from an opponent's doctor, he rapidly succumbed to delirious fever and the hostile climate, dying a victim of his patron's harsh demands.

This traditional version combines several appealing elements – the impoverished, wandering scholar, his unjust persecution for clinging to the truth, and the seduction of the mysterious icy North. Such accounts also reiterate misogynistic stories of Eve as temptress. From her frozen retreat, wealthy Christina lures Descartes, siren-like, to his death. She is literally a *femme fatale*, the ice-queen so chillingly portrayed by Greta Garbo. But there was another woman in Descartes's life – Princess Elisabeth of Bohemia. Focusing on Princess Elisabeth rather than on Queen Christina enables a different tale to be told. This view of the past not only changes the roles of women, but also suggests another way of remembering Descartes.

Despite her central European title, Elisabeth (1618–80) was the granddaughter of Bacon's witch-hunting James I and the aunt of England's George I; moreover, she lived in the Netherlands until she was nearly thirty. Soon after she was born, her parents accepted the crown of Bohemia, a foolish decision that cost them their fortune and condemned the family to a life of poverty. That is, poverty by royal standards – according to a running family joke, they dined on diamonds and pearls, the children's interpretation of pawning their inheritance.

Unlike her numerous brothers and sisters, Elisabeth was a sad, serious child, who resented her mother's gay life after her father's death. Seeking refuge in books, she was fluent in many languages. She was nicknamed '*la Grecque*' for her knowledge of Greek and Latin, but she also spoke the international French of royal families, the English and German of her parents, and the Flemish of the local people. (She later read Descartes's philosophy in the original Latin, translated English medical texts for him, and when she proposed that they study Machiavelli together, they both chose the Italian version.)[6]

It is hard to re-create the character and appearance of a princess, because all the contemporary letters and descriptions are couched in flowery, sycophantic language. Portraits are also unreliable witnesses – artists deliberately flattered their subjects, especially since profitable royal marriages were often confirmed on the basis of a picture rather than a meeting. Figure 14 shows a Dutch painting that is currently on display in the Bodleian Library in Oxford, where it hangs beneath a bearded Sophocles. It corroborates her younger sister Sophie's description of her as a tall, slim woman with 'black hair, a dazzling complexion, brown sparkling eyes, a well-shaped forehead, beautiful cherry lips . . .' With a tasselled hunting spear clasped in her right hand, Elisabeth is portrayed as Diana, the stern and athletic personification of chastity. Luxurious yellow and red feathers droop from her hat, and the diagonal yellow sash and blue dress accentuate her figure. Evidently not all the family pearls were lost, since Elisabeth (and also later Sophie) is painted wearing a necklace.[7]

Fig. 14
Elisabeth of Bohemia.
A portrait from the school
of Gerrit van Honthorst.

However faithful it may be, this portrait gives no hint of Elisabeth's embarrassment over her nose, which was long, thin and inclined to glow red. Like many talented young women, Elisabeth lacked confidence, and her sensitivity about her nose reinforced her timidity. Spurned by her mother, and mocked by the other children for her dreamy absent-mindedness, she peppered her letters with self-deprecating remarks about her appearance as well as her abilities. Elisabeth seems to have resigned herself to a studious life, recognising that impoverished, independent-minded princesses were not highly sought after as wives. Racked by anxieties about money and family, she was plagued by ill-health – lingering coughs, indigestion and many of the symptoms that we now associate with depression.

The winter of 1642: Elisabeth was twenty-four and Descartes was forty-six – almost twice her age, the same as her father would have been had he lived – when they met at one of her mother's parties in The Hague. Picture the scene: shy Elisabeth feels obliged to converse with this famous guest and express her admiration for his work; reclusive Descartes, conscious of his ageing, may have noticed her on a previous visit, but now confronts for perhaps the first time in his life a young woman who is genuinely more interested in philosophy than fashion.

In principle, Descartes endorsed female education, even claiming that he had simplified some of his explanations about God so that women could understand them. Elisabeth needed no such condescending interpretations. Possibly it was love at first sight; certainly their protracted friendship was deeply emotional. 'A body like those painters give to angels,' Descartes enthused in his first letter to her; she made him feel as if he had just arrived in heaven, he gushed. Theirs was an asymmetrical relationship, yet far more complex than its obvious, simplistic description – Elisabeth the adoring yet critical pupil, Descartes the older man who appeared besotted, yet came to shun her physical presence.[8]

Gossip-mongers were soon spreading rumours about her boat trips along the Rhine to Descartes's nearby home.[9] Although he stripped philosophical thought to its bare essentials, Descartes was no ascetic hermit. He lived in a spacious country mansion surrounded by orchards, slept late and employed an excellent cook. Little is known either about her visits to him or about the reciprocal, but infrequent, calls he paid on her family. Their subsequent correspondence suggests that they discussed intellectual matters. Perhaps Descartes taught her mathematics and encouraged her to participate in the experimental research he was pursuing – dissections and chemical experiments. As to other activities, one can only speculate. The letters that remain were composed with such decorum that they yield no firm evidence of a sexual affair, yet they do reveal a heartfelt warmth and intimacy on both sides.

Suddenly Descartes was gone. In May 1643 he abandoned his comfortable home and moved away to a remote village in the marshes. Did he resent his academic visitors and critics from the city, or was he fleeing from Elisabeth? Later that month, he accepted her polite rebuke about his failure to meet her, pointing out that he was often tongue-tied in her presence.[10] Although they rarely saw each other again, they regularly exchanged letters over the next seven years, right up to the time of his death in Sweden. Many of these letters survive – although there are tantalising gaps – and they reveal the extent to which Elisabeth influenced Descartes's thought and writings. Descartes's contemporaries thought his relationship with her was so important that in the first publication of his correspondence, his letters to Elisabeth were placed right at the beginning.[11]

Descartes was not the only scholar with whom Elisabeth conversed and corresponded, and at least two other writers dedicated their work to her. Her mother, the exiled queen, prided herself on her glittering intellectual circle, and other visitors included Margaret Cavendish's brother-in-law, who had taught her natural philosophy. As in the Cavendish household (see Figure 4), Elisabeth could engage in dinner-table debates with guests along with Sophie, who was also interested in philosophy (she later chose Leibniz rather than Descartes as her correspondent). In particular, Elisabeth was friendly with Anna van Schurman, eleven years her senior, who lectured at Leiden University and boldly campaigned for female education. Carefully cultivating her international reputation as the Dutch Minerva, van Schurman engaged in debates with several famous men, including Descartes. Elisabeth may well have been inspired by her example, although philosophically they stood on opposing sides. Van Schurman disapproved of Descartes's bids to revolutionise knowledge, and clung to older authorities – Aristotle and the Bible. The antagonism was mutual: Descartes broke off their relationship because he could not tolerate van Schurman's insistence on studying the Bible in Hebrew.[12]

Even in liberal Holland, Descartes was perceived as a dangerous radical in the first half of the seventeenth century. His enemies accused him of overthrowing orthodox teaching, demoting the Earth and its occupants from their central spot in the universe, and disproving God's existence. After his death, his writings were placed on the Catholic Index of Prohibited Books, and Cartesian philosophy was banned from French universities. The Cartesian, rational approach to the universe that lay at the heart of Enlightenment thought initially gained strength in the Netherlands, and only later spread to the rest of northern Europe. By choosing to ally herself with Descartes, Elisabeth was making a public declaration of her disagreement with conservative thinkers like van Schurman. Yet she maintained her own intellectual independence from Descartes – don't believe, she reprimanded him, that I agree with you from prejudice or laziness.[13]

At first, their correspondence was conducted on a master–pupil basis, heavily laced with elaborate compliments on both sides. But Elisabeth nudged the relationship ahead, refusing to let Descartes adopt an entirely didactic role. To his surprise, she solved a difficult mathematical problem he sent her; perhaps her unexpected success acted as some sort of quali- fying test, since the subsequent letters revolve around theological, philo- sophical and psychological issues. A staunch Protestant, she embarked on philosophical and theological exchanges with Descartes, a Catholic steeped in Jesuit principles. He felt stranded in a remote cultural desert, he told Elisabeth, and her searching enquiries were so unusual that it made him experience an extraordinary joy – he yearned to confess himself conquered.[14]

Descartes, the reserved scholar, confided in Elisabeth with unparalleled openness: his only extant reference to his mother appears in a letter to her. She reciprocated by consulting him about her most intimate experi- ences. Although, like many women, Elisabeth has left no systematic discus- sion of her ideas in a book, it is clear from her letters that she developed her own independent philosophy. Not afraid to criticise, she subjected Descartes's ideas to intense scrutiny.

Descartes's system sent women conflicting signals. His establishment of a detached, objective spectator reinforced his own superiority as the older, experienced philosopher. As a man, he wandered freely round Europe, while illness and convention confined Elisabeth to domestic self-observation. On the other hand, his emphasis on the power of an intellect detached from its body implies that a woman is just as capable of rational argument as a man. By the logic of his own philosophy, Elisabeth should be able to engage with Descartes as an equal. Her constant questions and criticisms forced Descartes to explain and also to modify his opinions, while her preoccupation with emotions, morality and her own health pushed Descartes in new intellectual directions.

The year after his abrupt departure, Descartes published his *Principles of Philosophy*, which he had been working on throughout the period he had known Elisabeth. Modern philosophers regard Descartes's *Meditations* as his greatest work, yet Descartes himself saw it only as a preliminary to his *Principles of Philosophy*, which was intended to be a systematic exposition of his entire life's thought.[15] And he dedicated this major book to Elisabeth, a remarkable testament of his high esteem. Dedications were normally ingratiating attempts to flatter a patron, but this one was different. Despite her title, Elisabeth had no money, and Descartes deliberately avoided the excessively ornate language adopted by authors hoping for financial backing. Instead, he wrote a short but scholarly and eloquent disquisition on virtue. Elisabeth's mind, he concluded, was unique: only she understood his ideas. Publicly declaring himself her 'most devoted admirer' who was captivated by her wisdom, majesty and gentleness, Descartes consecrated his book 'to the Wisdom which I perceive in you (because my Philosophy itself is nothing other than the study of wisdom)'.[16]

For Descartes, this philosophy of wisdom was not restricted to abstract ideas, but embraced everything to do with daily life. For modern readers, his *Principles of Philosophy* looks far more like a book about science than philosophy. Studded with diagrams, it discusses topics like mechanics, astronomy and magnetism. Like Newton, Descartes drew no hard lines between science, philosophy and theology, and it was only towards the end of the eighteenth century that these subjects started to split apart into separate disciplines. Newton became celebrated as the world's first scientist, while Descartes became the founding father of philosophy. By dividing up the scholarly territory, both England and France could boast their national heroes.

Three weeks after the *Principles* appeared, Elisabeth was ready to send Descartes her considered opinion. Naturally, she opened by professing herself unworthy of such an honour, declaring herself indebted to him for allowing her to share in his glory. The niceties over, she launched into some detailed criticisms, adopting a tactic that typifies her comments on his work. Unerringly placing her finger on a weak point, she disguised her acuity by claiming that the problem lay in her own weak intellect. 'I'm afraid,' ran the formula, 'that you will, with justification, withdraw your opinion of my ability, when you learn that I don't understand . . .' And then in for the kill. Like the huntress of her portrait (see Figure 14), she fooled her victim into a sense of false security before she pounced. Descartes tried to wriggle out of tight corners. Yes, he agreed, I didn't explain mercury's weight very thoroughly because I hadn't studied the metal properly, but my account was probably about right. The holes she picked in his magnetic theory, central showcase of Cartesian physics, were potentially even more threatening. After a few unconvincing sentences, he hastily drew the letter to a close: 'I humbly beg Your Highness to forgive me, if I only write in great confusion. I haven't got the book with me . . . and I'm constantly on the move.'[17]

Even in her very first letter, Elisabeth first elaborated her inferiority, and then rhetorically took advantage of this self-abasement to force Descartes into confronting his philosophical weaknesses. I've been hesitating to write, she told him, because I'm so ashamed of my disordered thoughts. But now I've plucked up my courage, she continued disingenuously, and I beg you to explain how a person's soul, which consists only of thinking matter, can cause the body to move. Writing with feigned ignorance, she had immediately targeted the central lacuna in Descartes's system. Only five days later he replied, graciously acknowledging the force of her criticism – your question is indeed, he admitted, the most sensible one to ask, and he set out to persuade her that his model worked.[18]

Elisabeth forced him to modify his earlier position. In her letters, she repeatedly prodded him, diplomatically masking her attacks as she pushed him further. Please excuse my stupidity, she protested with false naivety in her second letter, but your revised explanation still limps badly – you just haven't convinced me that a purely spiritual soul can move a person's body. You're right, he apologised, I didn't explain myself very well and I left some things out. And then he sketched out a new scheme incorporating a third mysterious substance that somehow unified the soul with the body. Descartes liked this philosophical sticking plaster so much that he included it in the *Principles of Philosophy*. But Elisabeth disagreed. Once again, she wrote, I'm embarrassed to send you further evidence of my ignorance, but . . . You yourself taught me about the importance of sceptical doubt, and I don't see how this hypothetical third substance works. Silence. No response. Elisabeth had neatly used a Cartesian argument to foil Descartes himself, and he apparently didn't like it (although, of course, the reply may have been lost).[19]

Wisdom, Descartes declared, included knowing how to stay healthy. Descartes the doctor is an unfamiliar figure, but he constantly dispensed medical advice to his friends, and informed the Duke of Newcastle (Margaret Cavendish's husband) that 'the preservation of health has always been the principal end of my studies'.[20] Under Elisabeth's influence, Descartes became increasingly interested in the links between mental and physical health, another aspect of the mind-body dualism so central to his system. At the outset of their correspondence, Elisabeth made Descartes swear the Hippocratic oath so that she could confide in him freely, and she sent him frequent reports on her condition. It seems to have been a gratifying arrangement for both parties: he provided a willing listener to her litany of complaints, while her gratitude and flattery led him to relish his role of paternal sage, the physician who would teach her how to cure her body with her soul. And when she was desperate, he obeyed her calls for help, travelling to her bedside to comfort her in person rather than through the mail.

Elisabeth was no passive recipient of Descartes's ideas. Under her guidance, their mutual emotional self-indulgence prompted Descartes to ponder more deeply on how feelings can produce physical symptoms. Faced with remedying Elisabeth's frailty, Descartes shifted his focus. Instead of worrying about the impact of health on psychological well-being and intellectual performance, he started to concentrate on using the mind to cure the body. Elisabeth taught him that she was burdened with a body that was 'very easily affected by the afflictions of the soul'; his injunction to rise nobly above her sickness was not, she protested, helpful for someone whose spleen and lungs had already succumbed to the physical effects of melancholy. Worrying about Elisabeth's indigestion led Descartes to agree: 'the soul undoubtedly has a strong effect on the body, as can be seen from the great changes caused by anger, fear and other emotions'. Elisabeth's outpourings induced him to formulate recipes for getting better: 'When the mind is full of joy, that greatly helps the body to feel good.' Before meeting Elisabeth, Descartes had prescribed medicine to make men wiser. Three years into their correspondence, he commented that the best way to live longer was to stop being afraid of death.[21]

Thought itself was therapeutic, advised Descartes. When Elisabeth plunged into a long period of decline in 1645, Descartes tried to rally her spirits. Ordinary people, he told her, succumb to their misfortunes, but superior beings – intellectuals like you and me, he meant – can make their reason overcome their emotions. Elisabeth's protracted periods of illness and despair often corresponded to major upheavals in her life – her brother's conversion to Catholicism, her own exile after being accused of inciting another brother to murder, and the beheading of her uncle, Charles I. Descartes may have been deeply sympathetic, but he did believe in the rational approach. Instead of conventional messages of consolation, he often sent cold analyses that read more like philosophical treatises. And this is, indeed, what they became. Descartes kept copies of his letters to Elisabeth, carefully filing them with other important documents. Their discussions formed the basis of his last book, *Passions of the Soul*, which explored – again – how the mind and the body interact.

A dose of Seneca, Descartes prescribed. No better way of restoring her spirits, he insisted, than studying the recommendations for happiness set out by this Roman Stoic. Without waiting for her acquiescence, a couple of weeks later he sent off a critique, the first in a series of letters rehearsing his arguments for his *Passions of the Soul*. Descartes was not simply analysing what Seneca had written. Instead, he was expounding what he thought Seneca *should* have said. Her opinion, he wrote, would both instruct him and help him to refine his commentary.

Elisabeth was not convinced by his interpretations. First came the self-demeaning flattery – please continue correcting Seneca, she begged, because your gifts of exposition make it all seem like common sense. Next were the reservations – however, I do have a small doubt . . . And then, straight to a fundamental problem. How can we reason our way to happiness, when so many misfortunes are beyond our control? If illness can take away our powers of thought, she asked, then how can the mind and body be completely distinct? And, she continued, since our health can influence our mental state, then the fact that I have a female body must surely affect how I think?[22]

Like Bacon, Descartes wanted to associate men with self-discipline and intellectual activity. Elisabeth undercut Descartes's arguments by pointing to the weakness of her own body. At first sight her debating strategy might seem strange, because it apparently reinforces arguments of female inferiority. Surviving correspondence reveals that this question plagued several female philosophers of the period. They worried about their tendency to blush and weep, uncontrollable behaviour that seemed to confirm their sensitivity to inner passions. But by stressing this physical vulnerability, learned women could emphasise their superlative intellectual powers: how much cleverer they must be than their male peers, with such obstacles to overcome! Mary Evelyn, wife of the famous diarist, was described as 'a great mistress of her passions' – high praise, suggesting that, like a man, she could control her emotions and exert her powers of thought.[23]

Back and forth went the letters, right through the winter. Elisabeth relentlessly hammered away, pointing out contradictions and forcing Descartes to explain his ideas, to redefine his terminology and make it more precise. For instance, under her urging, he acknowledged and spelt out the distinction between an infinite God and an infinite universe. Some of their head-on confrontations stemmed from their opposing religions. Descartes had claimed to base his entire system on certainty, yet he soon reintroduced God as an irrefutable escape route from sticky positions. For Elisabeth, the question of free will remained a major obstacle. If God is omnipotent, she demanded, how can people perform evil deeds that destroy happiness? Again, she used a Cartesian argument against Descartes himself – since God lets us feel free, she declared, therefore we effectively are. His back against the wall, Descartes fell back on vague terms like 'incomprehensible', 'of another nature'. In the end, he admitted, only faith can convince us of God's powers.[24]

Unconvinced by Descartes's attempts to explain how an immaterial soul might interact with a human body, Elisabeth defied him to define the passions. A few weeks later, in November 1645, he reported that he was still thinking how best to do this. The following spring, stimulated by their correspondence, he sent her a draught copy of the opening sections of his *Passions of the Soul*.[25] Now ignored except by specialists, this was his last book, the culmination of his life's thought. Advising Elisabeth how to make herself well provided the therapeutic basis of Descartes's prescriptions for living a good life, one that is ethical as well as enjoyable. How to be good and feel good – the most important question philosophers can ask.

For Descartes and his contemporaries, sadness, anger and the other emotions belonged to medicine, because they were associated with physical conditions. But they were also central to moral philosophy, which – Descartes insisted – was the ultimate wisdom because it demanded knowing all the other sciences. By persuading Descartes to focus on her sickness, Elisabeth had not diverted him on to a side track away from philosophy. On the contrary, she goaded him into clarifying issues about the relationship between the mind, the body and God that were central to his whole system. Swayed by Elisabeth, Descartes placed a new emphasis on personal experience and observation, moving his arguments from the purely metaphysical to a more physical plane.[26]

A few months after this sequence of letters came to a close, Descartes and Elisabeth met for the last time. On 15 August 1646, Elisabeth reluctantly left for Berlin, banished by her mother during the scandal surrounding the assassination of a young Frenchman, who was rumoured to have seduced both her mother and her sister. Perhaps Descartes suspected that they might never meet again – was he secretly relieved that she was moving far away? Certainly the tone of their letters became more intimate and relaxed, slightly gossipy even.

This time it was Elisabeth who nominated the intellectual tonic to revive her spirits – not Seneca, but Machiavelli, an appropriate diet for a princess victimised by political intrigues. She reassured Descartes of her success in applying his rational medicine to dispel her depression, and even managed to joke about the catarrh-ridden pedantic brains of her acquaintances. Yet despite avoiding a visit, Descartes continued to worry about her. This concern for Elisabeth provides one explanation for his decision to seek the support of Queen Christina via his close friend Hector-Pierre Chanut, the French Ambassador to Sweden. Descartes had never looked for patronage before, and he probably hoped that he could enlist Christina's protection for Elisabeth. As a first step, he sprinkled gratuitous flattering references to Elisabeth throughout his correspondence with Chanut. The lack of subtlety confirms his intentions: in his opening letter, without even pausing to start a new sentence, Descartes the ingratiating diplomat contrived to link the two women, praising their shared royalty that enabled them to 'surpass by a long way the learning and virtue of other men'.[27]

After almost three years of negotiations, Descartes finally set off for Sweden, having unfortunately missed the warship that Christina had sent to collect this intellectual trophy. Within a couple of days of his arrival, Descartes was – at least, according to him – singing Elisabeth's praises to Christina. Possibly attempting to boost Elisabeth's hopes, he emphasised Christina's generosity and virtue. Nevertheless, he confided, Christina seemed woefully ignorant of modern philosophy – he wasn't sure whether she would be dedicated enough to tear herself away from her old-fashioned studies of classical philology. Descartes omitted to mention that the Greek tutor was thirty years younger than him, but his scantily concealed desire to make Elisabeth jealous by praising Christina suggests that emotional as well as intellectual rivalries were involved.[28]

Elisabeth replied with sarcastic dignity. Don't think for an instant that your glowing description of Christina has made me jealous, she retorted. I'm happy to learn about such an accomplished woman, one who helps to remove the imputations of stupidity and weakness too often cast at women – although you do seem to have discovered even more marvels in the Queen than her reputation would suggest.

Elisabeth thanked Descartes for his kindness in bringing her to the Queen's attention. I'm happy, however, she concluded pointedly, that your high regard for Christina won't oblige you to remain in Sweden. Written more than two months before Descartes died, this letter must surely have reached him. But he didn't reply.[29]

Epilogue

Descartes was buried in unconsecrated ground, exhumed seventeen years later, and is now in his third Parisian resting place. Apart, that is, from his skull and his forefinger, whose whereabouts remain uncertain. His reputation has fluctuated. Since Descartes himself believed that the most important thing in life is 'to procure, as far as possible, the good of others', he would presumably have wished to be remembered for the moral guidance that he composed under Elisabeth's prompting. But this aspect of his work, together with his extensive research into optics, astronomy and many other scientific topics, is now forgotten. He is predominantly celebrated as a philosopher who dramatically altered the way we think, even though nobody agrees with him. As Voltaire put it: 'He was wrong, but at least it was with method.'[30]

Christina continued to study classical texts and ancient philology, just as Descartes had feared. Five years later, she created an international scandal by converting to Catholicism and abdicating. Descartes–Christina myths developed in two directions. Descartes's followers accused her of luring him to his death, even though there were other valid reasons for his journey to Sweden, including seeking patronage for Elisabeth. Reciprocally, Scandinavian critics blamed Descartes for his malign influence on Christina. Historians on both sides are trying to unravel these historical tales.[31]

Elisabeth immediately relinquished all hope of Swedish patronage. Judging from Chanut's response, she must have sent him a curt note demanding the return of her letters – no copies to be made, she specified (unsurprisingly, she refused to change her mind when Chanut informed her that Christina would find them interesting). Elisabeth, then aged thirty-one, apparently never discussed Descartes's death, but her relatives found her obvious depression hard to cope with. For the next seventeen years she lodged in a succession of family homes, coming to epitomise the unmarried and unwelcome older sister who can always be relied upon to help out at deaths and weddings. She then entered a Protestant convent, eventually becoming its head. At last this wandering princess achieved an identity of her own, ruling over a large household of aristocratic women and administering a territory of 7,000 people. Although she turned to religion rather than reason for consolation, she continued to discuss intellectual topics by letter. She corresponded with Leibniz, and also with philosophers interested in reconciling Christianity and Cartesianism. Provoking much scandal, she provided a refuge for her old friend and Descartes's opponent, Anna Van Schurman, who was being vilified for her attachment to a heretical minister. Elisabeth's sister Sophie reported that she calmly ordered her coffin, and then died like a guttering candle. A century later, her tomb was opened, and her silk robe crumbled to dust on contact with the air.[32]

Elisabeth may seem a solitary figure, but she was not a lone pioneer. She often encountered other scholars – women as well as men – and was bound in to the correspondence network stretching throughout Europe. In the middle of the eighteenth century, she was celebrated as the leader of a philosophical sect called the female Cartesians. Viewed in retrospect, these intellectual women formed an underground movement amongst Descartes's followers who criticised Cartesian ideas from within. In principle, if human beings are defined by their reason, then their bodies are irrelevant. So for the female Cartesians, Descartes's famous maxim *I think, therefore I am* could be converted into *I think, therefore I am ungendered*. However, this scattered, diverse group flourished only briefly, and female scholarship soon became suppressed in France. Ironically, Elisabeth and the other Cartesian women had helped to construct a masculine philosophy that came to exclude women. Using a new vocabulary, modern French feminists reiterate many of Elisabeth's original arguments against Descartes. But they are caught in the same trap as her – how can you fight rationalist thought with rational weapons?[33]

Chapter 4

Anne Conway / Gottfried Leibniz

Since I have the honour of writing to a Lady, I would not dare to discuss these matters at such an advanced level if I did not know how intelligent English Ladies are. I have seen an example in the work of the late Countess of Connaway, without mentioning others.

Letter from Gottfried Leibniz to Damaris Masham,
25 December 1703

Even male-genius versions of history can be rewritten – unsuspected heroes are discovered, and others lose their glory. At the beginning of the twentieth century, the philosopher Bertrand Russell was amazed to find that recent theories of mathematical logic had already been formulated 200 years ago and lain neglected. Russell's forgotten genius was Gottfried Leibniz, now world-famous as one of Germany's major mathematicians and philosophers, but previously edged to the historical margins. Leibniz served the Hanoverian court for more than forty years, yet none of his titled colleagues attended his funeral, and he was buried in an unmarked grave. During his lifetime, only a fraction of his myriad writings appeared in print. But the latest comprehensive edition of his papers currently takes up almost two yards of shelf space, and new volumes are still being added.

British scholars in particular used to be uninterested in Leibniz because he was one of Newton's most hostile rivals. But now, philosophers hail him as a neglected genius whose arguments about space, time and monads foreshadowed the new science of relativity. Historical fashions have also changed, and modern writers no longer ridicule his interest in alchemy and other activities that used to be dismissed as non-scientific. A modern summary of his life runs like this . . .

Leibniz was one of the last great polymaths, a man so well-read that he was known as a walking encyclopaedia. As well as making major innovations in mathematics, he dispensed advice on Chinese hexagrams, silver mining, fossils, legal education and religious reform. Although he was Germany's greatest philosopher, he profoundly disagreed with the two other intellectual giants of his age, Descartes (who died in 1650, when Leibniz was a toddler) and Newton, his English contemporary and rival. Newton and Leibniz both claimed to have invented calculus first, and they became engaged in a vitriolic correspondence debate. Their violent arguments were to do with international politics as much as mathematics, and they made a mockery of the ideal of scientific unity. Unfortunately for British mathematics, Newton's method was cumbersome, but continental mathematicians started to adapt Leibniz's ideas. It was only in the nineteenth century that Cambridge scientists successfully campaigned for the updated version of Leibniz's calculus, which lies at the basis of modern science.

Unlike Newton, Leibniz was not a university scholar, but worked in Hanover for the direct ancestors of the British royal family. He visited London twice, when he demonstrated a model of his new calculating machine that could multiply and divide, even find square roots, and he was unanimously elected a Fellow of the Royal Society. After his final employer – the Elector who became George I – acquired the British throne in 1714, Leibniz tried desperately to emigrate to England with him. But he was thwarted, partly because of Newton's machinations to curry favour with the new Hanoverian court in London. Leibniz was ordered to stay behind in Germany, where he spent the last two years of his life immersed in researching royal dynasties (sadly, he never got further than 1005 in his history of the Guelph family).

Leibniz's cosmological theories seem counter-intuitive. He argued that Descartes, Newton and the other atomists were wrong to believe that the universe is composed of individual, inert particles of matter. Instead, he insisted that the fundamental elements of nature are special entities called monads. Monads are hard to describe in ordinary words.

Unlike tiny atoms, they do not take up space. A better analogy would be to say that they resemble infinitesimal points of energy. Monads are important, Leibniz explained, because they are active, although some of them are far more energetic than others. It is only because some monads are relatively inactive that the world *appears* to contain inert matter. Living beings – their minds and souls, as well as their bodies – are colonies of monads that exist together in harmony, but never interact with one other.

Even philosophers find Leibniz's monads hard to understand. Imagine two clocks that seem to be linked together because they always tell the same time, but in fact are working perfectly, yet separately. Similarly, Leibniz's monads are in a sense preordained to give the appearance of collaborating harmoniously, even though they are operating independently. Leibniz used these concepts to prove the existence of God. In his time, theological, scientific and philosophical discussions were closely tied together, not split into separate academic disciplines as they were later. God, argued Leibniz, is the unique and perfect Creator of the universe, who has organised an infinite number of monads to make up the best possible world, one that approaches perfection. His was an optimistic, reassuring philosophy.

Leibniz's reputation has been resurrected, yet there are still other ways of rewriting his life. Historians have restored Leibniz to a more prominent position in the international one-sex philosophical fraternity, but some of his intellectual circles also included women. Through letters as well as in face-to-face meetings, Leibniz discussed science and philosophy with women throughout Europe. He also relied on female patrons for financial support, and paid tribute to women's insights. Like all relationships, these friendships were not symmetrical. They differed from one another, but were always conducted for the mutual benefit of both partners.

These women were indispensable to Leibniz's life. Although their intellectual significance is scarcely mentioned in standard accounts, they did affect the pattern of Western philosophy and how people think today. One of the philosophical women he most admired – Anne Conway – does not even appear in the index of Leibniz's standard English biography.[1]

* * *

One of Leibniz's correspondents was Elisabeth of Bohemia, who had engaged in an intensive philosophical debate with Descartes. But his closest and longest friendship was with Elisabeth's younger sister, Sophie, whose husband became the Elector of Hanover. Worried about his insecure position at the Hanoverian court, Leibniz diplomatically sought Sophie's allegiance. But their bond was far stronger and more complex than a straightforward bid for patronage. They exchanged more than 300 letters, which reveal not only her concern for his well-being, but also how he welcomed her opinions on his own ideas as well as on those of other philosophers.

As part of his job as the Elector's Librarian, Leibniz often travelled abroad to garner information for his history of the royal family – his stone of Sisyphus, he groaned, as the project started to take over his life. He wrote frequently to Sophie, keeping her informed about the latest gossip as well as new ideas in religion, medicine and philosophy, and she repeatedly questioned him on his monads. How could there be so many? How could they explain the differences between perishable bodies and immortal souls? Sensible questions – and he sent back long answers. It would be wrong to exaggerate Sophie's intellectual impact: clever as she was, she lacked the intellectual commitment of her sister Elisabeth. She made relatively brief philosophical comments, and her opinions neither pushed Leibniz in new directions nor persuaded him to change his mind. However, her constant probing did force him to focus on difficult points and explain his views more clearly.

Just as importantly, her backing consolidated Leibniz's position at court, and she fostered his international reputation by recommending him to her influential relatives. Leibniz diligently searched for evidence supporting Sophie's claims to the English throne. Eventually he was successful, but she died without ever realising that her son would become George I of England.[2]

Sophie's daughter, Sophie Charlotte, was another woman who played a key role in Leibniz's life. Prussia's first queen, she became internationally renowned as a learned woman who also enjoyed partying into the small hours. Sophie Charlotte was a hardier debater than her mother. Leibniz acknowledged how she sharpened his own thoughts. 'This great Princess,' he wrote, 'liked to be kept informed about my speculations, and she even deepened them.' Leibniz's failure to explain his ideas infuriated her, and she insisted that he write out long explanations. Her interest lay mainly in the religious aspects of Leibniz's work. How, she demanded to know, did his monads help to resolve the clash between a perfect God and the existence of evil?[3]

Taking advantage of her position as queen, Sophie Charlotte convinced her husband that a Prussian Academy, one to rival those of France and England, would enhance his international reputation. She ensured that Leibniz became its first President. It is, he wrote to her gratefully, 'the role of women of elevated mind more properly than men to cultivate knowledge'. Self-serving flattery, of course, but Leibniz did also genuinely support women who wanted to study. In his draft for the regulations, he declared that this new scientific Academy should encourage learning not only amongst the nobility, 'but among other people of high standing (including women)'. But although he seems to have recognised women's intellectual abilities, little changed in practice. The new Academy followed the London and Paris precedents, and women were effectively excluded until the twentieth century.[4]

Leibniz often stayed at Sophie Charlotte's palace, and they also exchanged many philosophical letters. Leibniz was devoted to this young woman who was his intellectual companion as well as his patron. But suddenly – pneumonia struck. Sophie Charlotte was only thirty-six when she died in 1705, long before her mother Sophie, Leibniz's other royal philosopher. A weeping attendant reported that her last words were of Leibniz. 'Do not feel sorry for me,' she whispered, 'for now I am going to satisfy my curiosity about the principles of things that Leibniz has never been able to explain to me: about space, the infinite, being, and nothingness'.

When she herself slipped into nothingness, Leibniz was grief-stricken. After weeks of despair, he gathered together the records of their long conversations. This may have started out as a project of self-therapy, but their talks became converted into his *Theodicy*, a set of essays that grappled with Sophie Charlotte's perpetual problem – how can a merciful God allow evil to exist? This was the only philosophical book that Leibniz published in his whole life, and it was the direct product of his debates with the Queen.[5]

* * *

Gradually a third royal woman gained Leibniz's allegiance – Caroline of Ansbach, who married George I's son (the future George II) and so became England's Princess of Wales in 1714. Despite her privileged birth, Caroline had had only a perfunctory education, but she dedicated herself to learning. Like Sophie Charlotte before her, Caroline was preoccupied with the theological and philosophical issues of evil that had entered into Leibniz's *Theodicy*. After she moved to London and struggled to become accepted amongst her new subjects, Caroline became renowned for entertaining an intellectual circle of England's finest scholars, including Newton, with whom she engaged in long conversations. Her correspondence grew to cover Newton's ideas about gravity, time and space, which were very different from those of Leibniz.

Stranded in Hanover, Leibniz had a royal ally right inside the English court – since George I's wife had remained behind in Germany, Caroline was Britain's most powerful royal woman. Leibniz relied on Caroline to ensure that his salary was paid, and urged her to help negotiate a position for him as court historian. As Newton and Leibniz jockeyed for favour with George I, Leibniz savagely attacked Newton in his letters to Caroline, attempting to discredit his rival and further his own interests. She participated in one of the most famous debates in scientific history – the triangular exchange of letters between Caroline, Leibniz and Samuel Clarke, a keen Newtonian propagandist. This correspondence was published in 1717 to advertise Newton's views, and did much to establish his still slightly uncertain position. Caroline was no mere box-office address. Although she is often depicted as only an intermediary, she argued with Clarke and Newton, commented on Leibniz's opinions and arbitrated in the arguments.[6]

Caroline was just one of many learned women in England whose contributions have been marginalised by subsequent historians. Another of Leibniz's correspondents was Damaris Masham, daughter of the famous Cambridge philosopher Ralph Cudworth and close friend of John Locke. Locke experts are at last acknowledging the important part that she played in the development of his ideas. In long letters – hers in English, his in French, the international court language – Leibniz and Masham debated philosophical issues over a couple of years. When she politely protested her unworthiness, he politely reassured her. The

Countess of Conway, he wrote, has already convinced me how brilliant English women are.[7]

Leibniz never met the Countess of Conway (1631–79), who was born Anne Finch and died fifteen years before he was introduced to her ideas. The following decade of his life was exceptionally fertile, the period when Leibniz consolidated a metaphysical system that bears strong resemblances to Conway's. Leibniz's references in letters confirm his deep admiration for her philosophy, as well as his expectation that other people should be familiar with her work. He evidently regarded her as an important intellectual predecessor.[8]

The international network of powerful women played a vital role in bringing about their interaction. The philosophical messenger who carried Conway's ideas from her large stately home in Warwickshire to Leibniz's rooms in Hanover was Francis Mercury van Helmont, renowned throughout Europe as a medical alchemist. He was the doctor of Elisabeth of Bohemia, who had originally sent him over to England to try to collect a pension for her. And how did van Helmont get to know Leibniz so well? Through Elisabeth's sister Sophie, who had forwarded two of his books to Leibniz, and discussed them with him in their philosophical correspondence.

Sophie invited van Helmont to Hanover for a few months in 1696, and every morning punctually at nine o'clock, Leibniz and van Helmont met in her rooms. Leibniz was surprised to find that this eminent physician, who was now a Baron, was dressed like a workman. Nevertheless, he sat and listened while van Helmont enthused about the new ideas he had developed together with Conway. 'He was a close friend of the Countess of Kennaway,' reported Leibniz (in slippery spelling), 'And he told me the story of that extraordinary Woman.'[9] This is an up-to-date version of what Leibniz learned.

As a child, Anne Finch adored her older brother John. Although he was often away at school and university, her love was reciprocated: he wrote her fond letters and sent her books to feed her avid intellectual appetite. Fluent in Latin and Greek, she read all the classical philosophers as well as the modern controversial authors – men like Descartes and Spinoza, who were accused of squeezing God out of their cosmology. Many clever girls must have studied vicariously in this way when their brothers went off to university. Others benefited from sympathetic fathers, husbands, or even perhaps their sons. For instance, John Evelyn first met Mary Browne when she was only thirteen, yet he was so impressed by her learning that he married her two years later and took over her education himself. She later spelt out how women could study through their men, ranking relationships in an interesting order when she described John Evelyn's different types of influence over her – 'a Father, a Lover, a Friend and Husband'. Her son's tutor lived in the house for six years, and she grasped the opportunity to carry out long intellectual conversations with him; when he left, they continued to correspond for the rest of her life.[10]

Conway's lifelong scholarly relationship was with her brother's philosophy tutor at Cambridge, Henry More. This eminent, charismatic teacher had introduced Descartes's ideas into England, even though he disagreed with some of them. More belonged to a group of Cambridge scholars, the Platonists, who emphasised the importance of religious, mystical experiences. Anne was about eighteen when John introduced them to each other, launching an intimate philosophical friendship that lasted until she died thirty years later.

In public, More openly declared his admiration for Conway's intellect, even dedicating one of his books to this 'noble Person . . . whose *Genius* I know to be so speculative, & *Wit* so penetrant, that in the knowledge of things as well Natural as Divine you have not onely outgone all of your own Sexe, but even of that other also'.[11] Their emotional relationship is harder to pin down. Surviving letters suggest that she was more devoted to him than to her wealthy husband, Edward Conway, who was often away from home. 'I profess, Madame,' wrote More, 'I never knew what belonged to the sweetness of friendship before I mett with so eminent an example of that virtue . . . But discretion bids me temper myself, and absteine from venturing too farr into so delicious a theme.'[12] Despite such tantalising hints, historians fall back on facile platitudes – since More was Cambridge's leading exponent of Platonism, his love for Conway is described as Platonic.

Conway spent most of her married life at Ragley Hall, a large country house near Stratford upon Avon. Now a tourist site owned by the Earl and Countess of Hertford, Ragley Hall is an imposing columned building which then lay at the centre of a large estate designed with geometrical regularity. Only one candidate for a portrait of Conway survives (Figure 15). It shows a solitary young woman absorbed in reading her letter, dwarfed by Corinthian columns and elaborate architecture. This picture, which

Fig. 15
Anne Conway's experience.
Samuel van Hoogstraten, *Young Woman with a Letter.*

partners a similar one of (probably) John Finch, belongs to a group of paintings by Samuel van Hoogstraten in which he played with perspectival illusions. Like other Dutch artists, van Hoogstraten enjoyed peering in on women holding letters, eliciting erotic sentiments through the mystery of the absent correspondent.

Even if this is not in fact Conway, the scene does evoke her daily existence, tucked away with her books in the Warwickshire countryside. Who might the letter be from? More, perhaps? Or her busy husband? And who is the shadowy man in the background? The seductive statues reinforce the intensity of her emotion, but the staring dog – conventional symbol for marital fidelity – challenges the gaze of us, the external viewers, as though the animal were warding off unwelcome visitors. Conway seems an open prisoner in this semi-enclosed courtyard, adorning her husband's mansion like the faithful dog and the cat trapped behind the railings.[13]

Conway's philosophical writing is scholarly but vivid, enriched by analogies drawn from her own daily life. She uses concrete examples, ones familiar to all her readers – freezing alcohol, digestion, rotting meat. Far from being idle ladies of leisure, women running large homes like Ragley Hall were skilled managers, responsible for the household's health as well as the preservation and preparation of food. Their expertise in gardening, brewing, herbal plants and curative medicines gave them a good knowledge of the biological and chemical sciences. When she wrote philosophy, Conway drew on her own experiences.[14]

Her most potent experience was pain. Ever since a childhood fever, she had been plagued by crippling headaches, which a succession of Europe's best doctors tried unsuccessfully to cure. This chronic illness made Conway a semi-invalid. Isolated at Ragley Hall, she saw her husband less often than doctors, faith healers and the devoted Henry More, who visited for months on end to discuss philosophy – intense conversations that affected the ideas of both. Together they studied modern controversial writers, including Descartes, Spinoza and Gassendi, as well as the older Jewish mystical works of the Kabbalah. Sometimes they were joined by Masham's father and other academic philosophers; and when More was away from Ragley, they continued their debates in long letters. Unlike Conway, More published prolifically, yet she effectively co-authored some of his books.

Together they developed a spiritualist, holistic cosmology very different from the one that Newton was about to introduce.[15]

More packed his letters with well-meaning advice on her health. Don't try mercury, he warned – instead, watch your diet, try this secret red powder from Wales. He also insisted on accompanying her to Paris for surgery. This proved an exciting trip. There were debates with the French doctors, who wanted to bore a hole in her skull to release the pressure of the vapours on her brain; they eventually decided to open her arteries instead. Even her husband's arrival was delayed because he was thrown into prison for a time. After the three of them had returned to England, it was More who eventually persuaded van Helmont to treat her, a decision he probably came to regret. The itinerant physician failed to cure Conway, but decided to stop wandering around Europe. Instead, he stayed at Ragley Hall for the last nine years of Conway's life, gradually pressing More away from his former privileged position.[16]

Conway's chronic pain also imbued her philosophical thought. While More was looking after Conway in Paris, he read Descartes's book on the passions and discussed it with her. Descartes's book had stemmed from his correspondence debates with Elisabeth of Bohemia, and – like Elisabeth – Conway realised from her illness that she disagreed with Descartes's separation of mind and matter. Conway proposed what we might call a more holistic approach. If the soul really was completely distinct from the body, she argued, then how could it 'suffer so with bodily pain? . . . why is it wounded or grieved when the body is wounded?' The solution was simple, she answered – just recognise 'that the soul is of one nature and substance with the body . . . the soul moves the body and suffers with it and through it'. Body and spirit, she believed, were composed of male and female principles, but she had faith in cooperation, not domination. In her version of physiology, these male and female principles functioned differently, yet collaborated.[17]

Medical men thought otherwise. As an aristocratic woman, Conway could summon England's finest doctors to Ragley Hall. Thomas Willis, an eminent anatomist from Oxford, made an intensive study of her illness, and incorporated his observations into his influential study of the brain and neurological disease. Hailed as a Baconian, Willis set out to penetrate nature, to peer inside human brains and 'unlock the secret places of Mans

Mind'. He retold the myth of Minerva's birth, casting himself as the man-midwife who delivered her from Zeus's brain with his instruments. By this 'Caesarean Birth,' he wrote, 'Truth will be brought to Light, or for ever lye hid.'[18]

Conway pointed to the blood as the essence of life. In contrast, Willis moved the human centre to the brain, and so made rationality – the attribute of masculinity – the body's dominating principle. The men clustered round Conway were convinced that she was ill because she was a woman. More boasted that he had staved off a fever by reading some mathematics, but this was definitely not a treatment he prescribed for Conway. Because she was a woman, he insisted, hard work would weaken her physically rather than helping her to rein in her passions. Her male advisers were genuinely concerned about her mysterious headaches, but they also felt threatened by her intellectual powers, which seemed more appropriate for a man than a woman. More worried that her brilliance would make men (him?) look inferior. 'Really Madame,' he wrote to her, 'I think it was designe in Providence to add the head-ache to all other gifts she bestowed on your Ladiship in mercy to our sex, that they might not be putt out of countenance and grow out of conceit with themselves by being so infinitely surpassed (in what they pretend to most) by yours.'[19]

Conway was caught in a double-bind. The more she studied, the more she was accused of risking her health because, as a woman, she was intrinsically incapable of intellectual labour. And if she stopped working? Well, then she would fall into the trap of confirming expectations that women were emotional creatures, very different from self-disciplined, rational men.

Leibniz freely acknowledged his female predecessor's importance, commenting that 'my philosophical views approach somewhat closely those of the late Countess Conway, and hold a middle position between Plato and Democritus . . . all things are full of life and consciousness'.[20]

Like Leibniz after her, Conway proposed that everything is made up from one single substance (although she did treat God and Christ as special cases). Mind and matter, Conway and Leibniz both believed, are just different combinations of various self-activating principles, which can develop and improve. (Put more technically, they were both monist vitalists.) Leibniz encountered Conway's version of monads just as he was incorporating this term into his own philosophy. 'Concrete matter,' she wrote, 'disperses into physical monads [and] is ready to resume its activity and become spirit.'[21]

Conway and Leibniz were both worried about God, and they both asked the same questions. If matter and spirit are distinct, they wondered, then how could a spiritual God have created the material world? And how could a benevolent God permit the existence of evil, or condemn His creatures to eternal hell? Together with More and van Helmont, as she explored these issues Conway deviated from Christian orthodoxy into the Kabbalah, the ancient Jewish texts. This move was less outlandish than it might seem – precepts derived from the Kabbalah also entered Leibniz's philosophy.[22]

Harmony, perfection, optimism: these themes pervade the views of Leibniz as well as of Conway. Drawing on the Kabbalah, Conway argued that God is good because He has given matter the innate ability to become perfect through its own efforts. Every creature moves from evil to good, she insisted, so that any individual's improvement was both physical and spiritual – from sickness to health, from sin to salvation. 'As we see from constant experience,' she wrote with memories of her own agony, 'through pain and suffering whatever grossness or crassness is contracted by the spirit or body is diminished.' Perhaps that was how she made sense of her ceaseless headaches – being ill was purifying her soul.[23]

In the last years of her life, Conway turned to Quakerism. More, who had faithfully stood by her for many years, was gradually crowded out of Ragley Hall by the attentive van Helmont and a stream of Quaker visitors. As Conway became sicker, she tried to emulate the Quakers' inner resilience, their resignation in the face of physical persecution. When she was forty-seven, her pain intensified to unprecedented levels, but her husband, inured to her constant sickness, coolly prolonged his trip round Ireland and was away when she died. Van Helmont thoughtfully preserved her body in spirits of wine so that her husband could take a last look before she was buried.

Overcoming their differences, More and van Helmont decided that the best way to commemorate 'that incomparable Person' was to publish her small notebook of pencilled jottings summarising her life's thought. So van Helmont took the manuscript off to Holland, the international publishing centre for controversial works, although eleven years went by before it appeared, first in Latin (perhaps translated by More), and then translated back into English as *The Principles of the Most Ancient and Modern Philosophy*. The original notebook has disappeared.

More wrote a long emotional preface that only surfaced later. He invited his readers 'to admire with me the Sound Judgement and Experience of this Excellent Personage'.[24] Conway's *Principles* never became widely known – but some of her ideas live on in Leibniz's philosophy.

Chapter 5

Émilie du Châtelet / Isaac Newton

Mme du Châtelet informs you sir that tonight at her desk, while scribbling some note about Newton, she felt a little summons. This little summons was a daughter who appeared immediately. She was laid down on a quarto tome of geometry.
Voltaire, *Letter to the Marquis d'Argenson*, 4 September 1749

Émilie du Châtelet, wrote Voltaire, 'was a great man whose only fault was being a woman'. Du Châtelet has paid the penalty for being a woman twice over. In her own lifetime, she fought for the education and the publishing opportunities that she craved. Since her death, she has been cast in the shadow of her lover, Voltaire, who excelled in self-promotion. 'I was destined for immortality from the time of my birth,' he boasted, and his reputation still distorts Enlightenment history.[1]

For more than two centuries du Châtelet (1706-49) has been cast as Voltaire's mistress, as though she were his possession or at best an intelligent secretary. Voltaire himself told a different story. He appreciated the effect of du Châtelet's presence on his own work, and praised her as a 'great & powerful Genius, the Minerva of France'. Figure 16, organised by Voltaire, illustrates how he acknowledged her scientific superiority. It is the frontispiece of a book about Newtonian philosophy. Du Châtelet and Voltaire worked on it together, but only his name is on the title-page. In the picture, du Châtelet hovers above Voltaire's head, casting Newton's divine wisdom down onto his hand – Voltaire is the scribe, and she is the knowledgeable interpreter.[2]

In this allegorical image, Newton sits suspended in the clouds, floating

Fig. 16
Émilie du Châtelet as Voltaire's Newtonian inspiration.
Frontispiece of Voltaire's *Élémens de la philosophie de Newton* (1738),
engraved by Jacob Folkema after Louis-Fabricius Dubourg.

like a saint between God and ordinary mortals. This sanctification started
in his own lifetime and has continued ever since. Like Descartes, Newton
has become such a heroic figure in science's past that history and legend
sometimes blend into one another. In conventional accounts of his life, it
can be hard to pick apart myth and reality. Newton and his achievements

have been tailored to fit the heroic model of scientific discovery. Alexander Pope's famous couplet, originally intended for Newton's tomb in Westminster Abbey, launched Newton's posthumous career as a legendary figure from England's distant past:

> Nature, and Nature's Laws lay hid in Night.
> God said, *Let Newton be!* and all was *Light*.[3]

In these lines, Pope presents Newton as a divine messenger who flicked the switch of Enlightenment illumination, thus immediately and single-handedly revealing the truths of nature that lesser beings had groped for unsuccessfully.

The falling apple story is another way of presenting the same concepts. Newton himself originated this myth of instantaneous insight, this tale of a Eureka moment when he (allegedly) realised in a flash of inspiration that the same laws that make an apple fall to the ground must also control the paths of the planets through the heavens. Was Newton really inspired by a falling apple? Other anecdotes about Newton that were once common knowledge have now been forgotten – such as discovering that his dog Diamond had upset a burning candle over yet another scientific master-piece and destroyed it for ever, or borrowing his fiancée's finger to tamp down the tobacco in his pipe. Because these have been rejected as fables, does it make the falling apple tale more or less likely?

Sound-bite versions of the past tell the same story more prosaically. A typical one-sentence encapsulation reads: *Isaac Newton was one of the greatest geniuses who ever lived, and he revolutionised the way that people thought about the world.* Here is a slightly fuller account of that interpretation . . .

Isaac Newton was born in 1642, the same year that Galileo died, when the Catholic Church still judged it heretical to believe that the Earth goes round the sun. By the time of his own death, in 1727, views about the universe had changed for ever, and scientific truth had defeated reli-gious obstinacy. His greatest book was his *Principia* (or *Mathematical Principles of Natural Philosophy*) of 1687. Written in Latin and geometry, Newton's *Principia* became a new scientific Bible. As well as setting out his three laws of motion, which form the basis of modern mechanics, Newton's

Principia described the force of gravity that holds the universe together. For the first time, one simple mathematical law could be used to describe events in the heavens as well as on the earth. Revolving planets, falling apples, bouncing billiard balls – from now on, a single neat formula wrapped up their behaviour and converted the chaos of nature into stable, predictable order.

Newton's other major work was the English *Opticks*, not published until 1704, but summarising a lifetime's research into telescopes, prisms and rainbows. After centuries of debate, Newton had demonstrated once and for all that ordinary sunlight is made up of seven basic colours. It was in *Opticks* that he stressed the importance of experiments, and he bequeathed clear, detailed instructions to his successors. He also included a long list of speculations that dominated scientific enquiry for the next hundred years, not just in physics, but in chemistry, biology and the social sciences as well.

Newton spent thirty years immured within Cambridge, absorbed in esoteric mathematical calculations and alchemical investigations, a lonely scholar whose work was too abstruse for all but the most brilliant to understand. But for another thirty years after that, he was the doyen of European science, holding court at his London home, presiding over the Royal Mint and the Royal Society. Voltaire crowned Newton the king of natural philosophy, the genius who had surpassed all his predecessors to rule supreme. 'Honoured by his compatriots,' reported Voltaire, 'he was buried like a king who had done well by his subjects . . . Very few people in London read Descartes, whose works, practically speaking, have become out of date.'[4]

Nothing in this potted biography is factually false, yet does it convey a true impression of Newton's life and influence? Even assessing the validity of apparently straightforward statements can be difficult, because many of them depend on opinion rather than fact. According to Voltaire and other Enlightenment ideologues, Newton represented the power of reason over religion. Yet how can this be compatible with Newton's own books, in which he gave God an indispensable role in the day-to-day running of the cosmos? The past is always open to revision, and older interpretations now seem misleading. Newton's ideas were *not* immediately accepted;

following his experimental instructions did *not* always yield the results he claimed. Instead of celebrating Newton as the first modern scientist, many historians now regard him as the last of the great alchemical magicians.

So how much faith can we put in Voltaire's claim that Newton toppled Descartes overnight? For one thing, this was a battle between two countries as well as between two cosmologies. Enlightenment natural philosophers boasted about their love of truth, but the intellectual world was riven with international rivalries, which were just as bitter as political hostilities. As an English mathematician commiserated with a French colleague, 'We have our Newton, the Germans their Leibniz, and you your Descartes.'[5] Voltaire had his own axe to grind. After a period in exile, he pursued his agenda of taunting the French authorities by lauding England and denigrating France. By deliberately exaggerating Newton's success, Voltaire could criticise French culture.

Newton did eventually oust Descartes in France, but only after a long propaganda campaign waged by his supporters. One person who played a key role in bringing Newton before the French reading public was Émilie du Châtelet, a superbly energetic woman who committed herself to academic study with almost obsessive fervour. 'Ambition is an insatiable passion,' she reflected, and she made herself into a leading expert on Newton. The birth of du Châtelet's fourth child killed her, yet by then she was internationally renowned for her scientific knowledge.

Du Châtelet had an impressive range of expertise. Her annotated translation of Newton's Latin *Principia* remains the only complete version in French. As well as her scholarly text synthesising Europe's three rival philosophers (see Figure 12), she collaborated in producing two simpler accounts of Newton's ideas which helped to displace Descartes's supremacy. In addition, she wrote and translated works on other subjects – the nature of fire, happiness, the Bible, Greek poetry, morality. No wonder that she was praised in print for 'holding an advantage over Newton himself: she united the depth of Philosophy with the most acute & delicate taste for Literature'.[6]

'Judge me for my own merits,' protested du Châtelet; 'do not look upon me as a mere appendage to this great general or that renowned scholar.' Figure 16 shows her interposed between two famous men, Voltaire and Newton, as if she were a medium of communication between them.

Listening to her plea for recognition in her own right entails giving her an independent existence, uneclipsed by the shadows of these two Enlightenment geniuses. But how faithfully can this be done? Like other women of this period, she seems to behave inconsistently, and gives contradictory accounts of herself. Sometimes du Châtelet sounds as if she had reconciled herself to a secondary role as translator and interpreter, as though she sited herself in Newton's shadow. 'God has refused me any kind of genius,' she confided to one correspondent; 'I spend my time unravelling truths discovered by other people.'[7]

In some ways her behaviour conformed to stereotypical expectations for an aristocratic hostess: she loved clothes, dancing and entertaining. On the other hand, this apparent resignation to her feminine status clashed with her dedication to Newtonian natural philosophy, which meant constantly transgressing the norms for her society. Du Châtelet did engage in unconventional activities, yet at the same time she remained confined by traditional demands. Some of these she imposed on herself; in other cases, she seems to yield to pressure from her contemporaries. As if conditioned from birth into a form of psychological captivity, even in the midst of achievement she self-deprecatingly judged herself to possess only a second-rate talent.[8]

Faced by overt exclusion as well as ingrained doubts about her own capacity, du Châtelet was trapped between conflicting, unsatisfactory stereotypes – the learned eccentric, the flamboyant lover, the devoted mother. Even now, biographers plump for one or another of these hackneyed models. 'I am in my own right a whole person,' she insisted.[9] Hopefully she would have appreciated this version of her life.

Émilie du Châtelet was tall and beautiful. Many intellectual women would object to an account that started with their looks, but du Châtelet was extremely concerned about her appearance. She was proud of being included in the frontispiece of a book on Newtonian philosophy (Figure 17), which shows her dressed in an elegant low-cut gown, strolling in some palace grounds. She is discussing Newton's ideas with her Venetian friend Francesco Algarotti, who had written the book while he was staying at her country mansion. 'Did you recognise me in the picture?' she boasted to Pierre Maupertuis, her close friend and mathematics teacher. Du

Châtelet spent a fortune on her clothes and jewellery, acquiring the money from her husband, a succession of lovers, and card games in which her mathematical skills made her an impressive opponent – although she was also a heavy loser.[10]

Her success at court and with men made her an easy target for jealous rivals: 'Imagine a tall, dry woman . . . with a narrow face, a pointed nose, a dark flushed complexion . . . and there you have the face of the fair Émilie, a face with which she is so pleased that she spares no effort to display it: curls, top-knots, stone jewellery, glass jewellery, everything in great profusion.'[11] Malicious, yes, but hardly the description of a frumpy scholar. Du Châtelet lived at a high pitch, and devoted the same intensity to gambling, dressing and loving as she did to learning. When deadlines were close, she allegedly scarcely slept, plunging her hands into ice-cold water to keep herself awake. The major goal of life, she believed,

Fig. 17
Émilie du Châtelet discussing Newton with Francesco Algarotti.
Frontispiece of Francesco Algarotti's *Il Newtonianismo per le dame* (Naples, 1737), engraved by Marco Pitteri after Giambattista Piazetta.

was to be happy, and she gained the reputation for living enthusiastically. One should, she taught, indulge, yet also balance one's passions for food, sex and work.

Obviously brilliant as a child, du Châtelet resented the discrimination that made it impossible for her to pursue the same path through life as a man. On the other hand, unlike now, there were no conventional career tracks to follow even for a man who wanted to study science. Like women, men had to carve out their own routes to success, especially if they were not born into rich families. Protecting her own interests was important for du Châtelet. Women, she wrote, need to foster their own happiness, and so should study 'to console them for everything which makes them dependent on men'.[12]

Born into a wealthy aristocratic family, du Châtelet (née Gabrielle Émilie Le Tonnelier de Breteuil) benefited from an enlightened father. Instead of sending her to a convent school, he decided that she should be taught at home, and – like Elisabeth of Bohemia – she received the sort of education that was more typical for boys than for girls. She could apparently speak six languages when she was only twelve years old, including English as well as Italian. Languages, modern literature, the classics – she excelled in them all. Free to browse in her father's well-stocked library, she displayed such precocity that her father enjoyed showing her off to his intellectual friends.

As with young male prodigies, du Châtelet's learning was almost exclusively in the humanities. She encountered the philosophies of Leibniz and Descartes at seventeen, but it was only ten years later that she started to immerse herself in mathematics and Newtonian philosophy. By then, she was married to an older army officer, the Marquis du Chastellet (Voltaire later revised the spelling), had given birth to three children (two of whom survived) and was simultaneously developing friendships with two other men, Voltaire and Pierre Maupertuis. Maupertuis was a controversial young member of the Paris Academy who had just published France's first book on astronomy to be based on Newton's ideas. Because of the strong Jesuit influence, Cartesian philosophy still reigned, and Newtonian natural philosophy was not taught in the schools and universities until the second half of the century. She persuaded Maupertuis to teach her mathematics, although she often accused him of not taking her seriously enough.

By studying Newton, du Châtelet allied herself with a radical faction, and so was making a political as well as a scientific decision.

Much to her frustration, as a woman, she was excluded from two places in Paris where Newton was discussed: the Wednesday meetings of the Royal Academy and – less formally – Gradot's café. Paradoxically, this woman who danced and gambled in Parisian palaces led an isolated life, mocked as a freakish woman yet at the same time unable to enter male academic circles. She was trapped between the sexes. The German philosopher Immanuel Kant respected her ideas, yet felt that she must be some sort of hybrid creature: 'a woman who . . . conducts learned controversies on mechanics like the Marquise de Chatelier might as well have a beard.'[13]

Assessing du Châtelet's character is difficult. For one thing, she self-protectively concealed herself within deceptive outer shells. 'Émilie puts so much effort into appearing what she is not,' quipped one of her acquaintances, 'that no one knows any longer who she really is.'[14] To add to the problem, many reports are biased. At the time, snide gossips almost certainly exaggerated any strangeness in her behaviour, and distorted incidents that may or may not have happened. Even modern biographies are not necessarily reliable. According to them, du Châtelet resorted to dressing as a man so that she could drink coffee at Gradot's with Maupertuis and his friends. This anecdote has been repeated so often that it has acquired the aura of truth, yet no original source has ever been given. Historians sometimes sneer that her attachment to mathematics was a mere flirtation, a ruse to gain the attention of scholars who became her lovers. Perhaps it was initially – being clever doesn't preclude being beautiful and sexy. But what is certain, and far more important, is that she dedicated herself to a gruelling programme of mathematical study with Maupertuis and other tutors, including his close friend Aléxis-Claude Clairaut.

It remains uncertain whether she had an affair with Maupertuis, but there is no doubt about her relationship with Voltaire – they fell passionately in love. Voltaire was in serious trouble: his book praising England and criticising France had been banned, and a warrant was out for his arrest. Du Châtelet sent him off to Cirey, her run-down estate in the country that had formed part of her marriage arrangements. 'My Hermitage' she called it – and Cirey was conveniently near the border so that Voltaire could escape if the authorities managed to track him down.[15] While Voltaire started to pour money into renovations, she stayed in Paris, where – as well as studying and enjoying herself – she tried to restore his reputation.

In October 1734, du Châtelet joined Voltaire in Cirey for a couple of months. After making his alterations, and spending more of his money, she retreated to Paris, Maupertuis and the card tables. The following March, Voltaire was awarded his freedom, and he presented du Châtelet with an ultimatum: live separately in Paris, or together in rural seclusion at Cirey. She made up her mind. By the summer, Voltaire and du Châtelet were ensconced together at Cirey, mutually embroiled in a private world of intense intellectual activity entwined with romance.

The first step was to buy books for their libraries and instruments for their experiments. Together they eventually amassed 21,000 books, more than in most European universities, and they each had separate rooms packed with equipment. Du Châtelet took over the great hall, where she tested Newton's theories with wooden balls swinging from the rafters and metal apparatus forged from the nearby iron mines. She also had her own study, where the desk groaned under instruments, books and mathematical manuscripts. She decorated her chambers in blue and yellow, even coordinating the dog's basket into her colour scheme. The lavish fittings included several large mirrors, a huge walk-in wardrobe to accommodate her clothes, and a triple-locked jewellery closet. A secret passage led to Voltaire's suite of rooms – a small oak-panelled antechamber and a long gallery overlooking the gardens. In addition to his valuable ornaments, three pieces of furniture were placed against the end wall: a bookcase, a glass-fronted cupboard packed with scientific instruments and a heating stove disguised as a large statue of Love.

Like many couples, during their fifteen years at Cirey they developed their own daily schedules, so that they could spend time apart as well as together. Du Châtelet liked to rise at dawn while Voltaire was asleep, so that she could organise the household, deal with correspondence and see her children. Then she went upstairs to work, reportedly elaborately dressed and coiffed, until dinner – a very movable feast, as hungry guests discovered. Dinners were long, extravagant affairs, served through trapdoors so that the servants would not interrupt the intense debates.

Sometimes du Châtelet and Voltaire studied together, observing each other's experiments, discussing their latest ideas and annotating each other's manuscripts. But du Châtelet also liked to work alone in her room on her private projects. One summer Voltaire became absorbed in research for an essay competition about the nature of fire. After du Châtelet carried out her own investigations, she decided that Voltaire – who relied heavily on Aristotle – had reached the wrong conclusions. Only a month before the closing date, she decided to compete herself, but kept her plans secret in order not to annoy him. After many sleepless nights, she submitted her entry – an erudite memoir of 139 pages. She felt that since her essay would be submitted without her name, it stood a better chance of being fairly judged. 'I wanted to test my strength under the cover of anonymity,' she wrote to Maupertuis.[16]

Neither Voltaire nor du Châtelet won the prize (which went to some Cartesians), but he later persuaded the Academy of Sciences to publish both their entries – the poet/lover combination made good publicity. Knowing that her essay would be appearing under her own name, du Châtelet meticulously checked the details to protect her reputation, and became engaged in a head-on confrontation with one of the Academy's leading natural philosophers, who took her arguments seriously enough to oppose her publicly in print.

Du Châtelet and Voltaire welcomed visitors, but did not permit them to disrupt the daily schedule of study. Some of the intimate details of life at Cirey have survived in the letters of gossipy correspondents, although they can hardly be totally reliable – it must have been enormously tempting to twist a relatively innocuous event into something more dramatic for entertaining a distant reader. Her husband was often away at war, but he soon adapted himself to the unconventional ménage, coming to stay and also publicising their manuscripts. In addition, to compensate for being excluded from more orthodox academic circles, du Châtelet invited France's leading scholars to join her intellectual court. Several eminent mathematicians and natural philosophers came to Cirey, including a hired tutor who stayed for two years, Maupertuis (who became great friends with Voltaire) and Clairaut, who was heralded as France's new Newton.

* * *

During their second summer at Cirey, du Châtelet read Italian classics and studied natural philosophy with Voltaire. By the time of Algarotti's visit, Voltaire recalled later, 'he found her sufficiently skilled in his own language as well as familiar with the works of Newton to give him some very excellent information from which he profited'. Algarotti was impressed by Newton. 'Who would ever have believed,' he remarked, 'that England, which was reputed a country of dolts, should so excel and give laws in the sciences?' Algarotti had studied at Bologna, a progressive city where women's education was encouraged. Even one of the university's professors, an expert on Newton, was a woman – Laura Bassi, the envy of learned women all over Europe. By the standards of the time, Catholic Italy permitted girls to study a wide range of subjects. Bassi was afforded the status denied to du Châtelet, and other ambitious Italian girls were following her example by taking courses in natural philosophy.[17]

Algarotti completed his *Newtonianism for the Ladies* while he was staying at Cirey, and it first appeared, in Italian, in 1737. Like many educational books of that period, it is written as a dialogue between the author-teacher and the reader-pupil. Algarotti portrays himself instructing an imaginary Marquise who was glamorous, flirtatious and not very bright. The text is packed with sexual innuendoes and references to female vanity. Taken at face value, this slim book seems a laudable attempt to extend scientific education towards women. Viewed more sceptically, it undermines female equality by reinforcing the notion that women are inherently incapable of serious thought. Mathematics, thought Algarotti, was certainly beyond their capacity. The frontispiece (Figure 17) showed the refined young man – a swan, sneered Voltaire – with a real-life Marquise, his hostess and Newtonian colleague, du Châtelet. Before publication, she had been thrilled; afterwards, she was angry at being linked with his flighty heroine. 'His book is frivolous,' she fumed, and accused him of aiming at the boudoir market.[18]

The following year, 1738, a second Cirey book on Newton appeared, this time in French: *Elements of Newton's Philosophy*, with Voltaire's name on the title-page. Voltaire had smuggled it out to Holland for publication, but had found time to insert some last-minute jibes about his Italian rival's flowery style and to write a long poem explicitly addressing the book to a non-imaginary Marquise – Émilie du Châtelet. Far more solid than

Algarotti's superficial sketch, this *Newton* was well illustrated and clearly explained the basic principles of Newton's discoveries in mathematical astronomy and optics. For the first time, Newtonian ideas were accessible to a wide range of French people, and this book was a huge success.

The publication of *Elements* is often seen as a turning point in French science. According to many older accounts, Voltaire single-handedly convinced the nation of Newton's superiority to Descartes. More convincing versions describe a slower rate of change and point out that Voltaire seized a particularly opportune moment to publish. This was exactly the right time to market an introductory text because French people were already half-converted to Newtonianism. Maupertuis had been promoting Newton (and himself) by sending back dramatic reports of his exploits in Lapland, where he led an expedition to settle a crucial question about the shape of the Earth. According to Descartes, it was elongated like a lemon; according to Newton, it was slightly flattened at the poles. Maupertuis, just as adept at self-publicity as Voltaire, converted the ambiguous results into an outstanding victory for Newton – and also for Maupertuis, who became France's new scientific hero.[19]

As another blow to the traditional story, although *Elements of Newton's Philosophy* came out under Voltaire's name, it was a joint production. The frontispiece (Figure 16) shows Voltaire sporting his poet's laurel wreath, assiduously transcribing the words of his female muse – 'Minerva dictated and I wrote,' he told a friend. Du Châtelet is holding a mirror, a doubly appropriate symbol that immediately identifies her as the goddess of truth, but also alludes to her interest in her appearance. To reinforce the message, Voltaire included a foreword that made his debt to her clear: 'The solid study that you have made of several new truths and the fruit of considerable work, are what I am offering to the Public for your glory.'

It is impossible to reconstruct their individual contributions precisely, especially as the book was extensively edited in Holland, but du Châtelet's involvement was substantial. And it was she, not he, who sent an abstract to one of France's top scientific journals. To prepare himself, Voltaire plunged into Newton's *Opticks* – unlike du Châtelet, he could not cope with the complicated mathematics of Newton's *Principia*. As Voltaire bought himself more and more equipment – a telescope, an air-pump, thermometers and barometers – he confided to a friend that 'in all of this, Madame du Châtelet is my guide and my oracle'. Just as Voltaire's private letters refer to their collaborative study of the 'Émilienne philosophy', so too he was not embarrassed to declare his intellectual allegiance publicly. 'Let me stand next to you,' proclaims his florid verse preface, 'and show Truth to the French nation.'[20]

Early 1737: at last du Châtelet had some time to think. The house renovation work was more or less complete, Algarotti had been and gone, Voltaire no longer needed her help with their *Newton*. While they studied together, she had been carving out her own position, and now she set off in a new direction. Voltaire, who hero-worshipped Newton and was weaker than her mathematically, often accepted what other scholars told him, whereas du Châtelet was an astute critic who formulated her own ideas. When a new edition of *Elements of Newton's Philosophy* came out ten years later, Voltaire saluted her: 'I used to teach myself with you. But now you have flown up where I can no longer follow.' Du Châtelet's own two scholarly books about Newton grappled with the philosophical relationships between the Cartesian, Leibnizian and Newtonian systems.[21]

Unlike Voltaire, du Châtelet believed that good science demands metaphysical foundations. She was not satisfied with knowing *how* the universe works: she wanted a rational explanation of *why* it works. The nature of matter, the role of God, good and evil – these were the sort of issues that du Châtelet worried about. She accused Descartes of providing answers that were wrong; she criticised Newton and Voltaire for avoiding the questions. Newtonian science, she pointed out, purports to be based solely on experimental observations, yet inevitably entails metaphysical assumptions about the existence of scientific laws.[22]

For about a year and a half she prepared a new manuscript, *Foundations of Physics,* in which she tried to integrate the ideas of Descartes, Leibniz and Newton (see Figure 12). She wrote in complete secrecy, trapped in a dilemma. She desperately needed constructive criticism, but risked painful mockery by revealing that she, a woman, was engaged in such innovative work. Many learned women experienced similar conflicts. Almost a hundred years later, the English mathematical physicist Mary Somerville was working on a mirror-image project – elucidating the cosmological ideas of 'the French Newton', Pierre Laplace, for an English audience. 'I hid my papers as soon as the bell announced a visitor,' Somerville confessed, 'lest anyone should discover my secret.' Even close friends remained unaware that she was producing a book that would modernise university teaching and make her famous.[23]

At the last moment, du Châtelet panicked. Several chapters had already been printed, but she withdrew the book, embarked on an intensive course of mathematics with a private tutor and rewrote large sections to incorporate Leibniz's views. Perhaps – it seems a likely step – she had decided to confide in Maupertuis, who was fresh back from Lapland. But now she faced another nightmare: rumours spread that she was plagiarising her hired tutor's work. Vainly hoping to quell the gossip, she rushed into print.

In welding together three conflicting systems, du Châtelet suggested a novel yet welcome approach. Helped by the *Newton* that du Châtelet and Voltaire had worked on together, Newtonian ideas were spreading through Europe, but there were some fundamental clashes between the ideas of Europe's three eminent natural philosophers. In the universes of Newton and Descartes, mechanical forces operate on passive matter. By incorporating Leibnizian ideas about active substances, du Châtelet helped to forge a new version of Newtonian physics that resolved the problems of describing force and movement. But not everyone accepted this Leibnizian intrusion, and she became involved in a long-running international controversy that eventually resulted in new physical laws of energy. Her arguments were known and discussed: her public dispute with the eminent Secretary of the Paris Academy was analysed by Kant in his first published work.

Du Châtelet's *Foundations of Physics* was well received. It was reviewed in prestigious journals, and Maupertuis promoted it – and her – lavishly, writing that her book would have done credit to a leading member of Europe's prestigious academies. Unlike some commentators, he complimented her extraordinary achievement as a woman, yet without patronising her. The author's brilliance, he observed, is surpassed only by her modesty, since she chose to publish anonymously. After her identity had been revealed, du Châtelet was elected to the Bologna Academy – but never to the Paris Academy of Sciences.

In the years after *Foundations of Physics* finally appeared, du Châtelet continued to experiment and to study, but she kept getting side-tracked.

Her days were not entirely under her own control. She spent months on end in Brussels, overseeing a protracted legal case about her husband's property, and moaned about the time she devoted to obtaining a military commission for her son. Voltaire was a time-consuming partner who demanded editorial assistance as well as emotional care, and they spent a lot of time travelling, sometimes to escape the police. Although she pestered her aristocratic friends in Paris to back him, Voltaire's defiantly anti-establishment writing antagonised many people. And as well as running several households, du Châtelet got swept up in other exciting activities – dinner parties, gambling, amateur theatricals.

Like many women, du Châtelet turned to translation, intellectual work that tolerates frequent interruptions. She had already translated literary works, and now she was absorbed in natural philosophy. Good translations are vital for spreading new ideas, and modern international science could not have developed without them. For du Châtelet, translation meant more than just converting words into another language. She believed she should create her own version of a text, one that incorporated her own ideas as well as those of the original author. Newton had written his *Principia* in Latin, and had used complicated arguments; even the English version had errors. Du Châtelet aimed to give academics a scholarly study that interpreted Newton's ideas as well as translating his words into Europe's international language – French.[24]

Her work may have been intermittent, but it was thorough. To start with, she studied Newton's three different editions of the *Principia*, as well as his further attempt (published separately) to explain how his mathematics described the cosmos. Then there were all the commentaries, by English and Dutch disciples as well as by critics. And on top of that, French and Swiss mathematicians had developed his theories algebraically. She corresponded with experts, especially two who were also Cirey visitors: François Jacquier, a Jesuit priest and physics professor who lived in a monk's cell decorated with a *trompe l'œil* portrait of Newton; and Clairaut, her long-standing mathematics tutor, who had travelled to Lapland with Maupertuis and was now France's leading authority on Newtonian astronomy.[25]

By 1745, she had started to translate, snatching the early hours of the morning to work in peace. Clairaut was so impressed by the accuracy of her thought and the clarity of her language that he recommended her work to the royal censors, whose approval was necessary for all books published in France. But the whole project got delayed as she immersed herself deeper and deeper. Du Châtelet decided to undertake not only the literal translation of the text itself, but also three further types of translation. For newcomers, she converted the complex mathematics into elegant prose, supplemented by her own examples. Next, she turned to calculus, translating Newton's geometry into the new continental algebra. And finally, she summarised recent mathematical research and experimental vindications of Newton's theories.

Like Maupertuis, Clairaut was adept at self-promotion. In 1747, he made a dramatic announcement to the Paris Academy – Newton's law of gravitation was wrong! Newton had undertaken the complicated calculations involved in working out how the moon moves under the simultaneous tug of the Earth and the sun. This was, he said, the only problem that ever gave him a headache. Now, twenty years after Newton's death, Clairaut claimed that the moon was not where it should be according to Newton's physics. While everyone argued, Clairaut spent eighteenth months checking his figures, and then emerged theatrically to declare that Newton was right after all. Both Clairaut and Newton gained publicity from this manoeuvre.

Meanwhile, a domestic drama was being played out at Cirey. To her surprise, du Châtelet discovered that she was pregnant. Then aged forty-three, she was an elderly woman by contemporary standards. Although Voltaire was not the father, he helped her deceive du Châtelet's husband into thinking that the baby was legitimate. Plagued by gloomy premonitions, du Châtelet intensified her work schedule, working eighteen hours a day to finish in time. And she did finish, with a couple of weeks to spare. But her predictions of mortality were correct. A few days after her daughter was born, she fell ill. On her final day, 10 September 1749, du Châtelet asked for her Newton commentary and recorded the date on it. Then, lapsing into a coma, she died.

As obituarists vied to produce witty digs disguised inside flowery compliments, du Châtelet's manuscript mysteriously vanished. Nevertheless, someone evidently had a talent for timing, because ten years later her *Principia* was – like the earlier *Elements of Newton's Philosophy* – published at a new peak of public interest in Newton. Clairaut had just helped to achieve yet another victory for Newtonian astronomy. For decades, French astronomers had been arguing about Edmond Halley's forecast that the 1682 comet would return in 1758–9. Using the latest mathematical techniques, Clairaut recalculated the date, but made his propaganda more effective by refusing to divulge his methods. Everyone waited to see if he was right.

And he was: the comet arrived on schedule, the newspapers heralded Clairaut as the new Newton, and du Châtelet's sparkling new translation appeared in the book shops. A publicity machine for France's 'illustrious female scholar' moved into action. 'The public has been waiting impatiently for years,' enthused one journal, insisting that the unexplained delay made the book 'even more stunning, by contributing to philosophy's moment of triumph'.[26] This advertisement of Newton's victory was overconfident. Extraordinary as it now seems, there were fierce debates in France during the 1770s about a French experiment that claimed to disprove gravity: objects were apparently getting heavier rather than lighter as they were moved up a mountain away from the centre of the Earth. Although this was an elaborate hoax, its success illustrates how powerfully and for how long anti-Newtonian lobbies operated in France.[27]

It took the best part of a century for Newton to displace Descartes and become France's new God of Reason. Du Châtelet played a vital role, because she explained Newton's ideas with unsurpassed clarity, and also made them more palatable by integrating them with other philosophical systems. The *Encyclopédie*, itself packed with propaganda for Newton, praised seven writers who had made Newtonianism easier to understand. Only one of them lived in France – Émilie du Châtelet.

Domestic Science

You are the true Hiena's, *that allure us with the fairness of your skins; and when folly hath brought us within your reach, you leap upon us and devour us. You are the traiters to Wisdom; the impediment to Industry . . . the Fools Paradise, the Wisemans Plague, and the grand Error of Nature.*

Walter Charleton, *The Ephesian Matron,* 1668

Idling away the hours at home, Euphrosyne longed for the vacations, when her brother Cleonicus came back from university. Then she could put aside her sewing and painting, and indulge herself in far more interesting activities – learning how to use telescopes, microscopes and air-pumps, or experimenting with the latest invention, electrical machines (Figure 18). 'I often wish it did not look quite so masculine for a Woman to talk of Philosophy in Company,' sighed Euphrosyne; 'how happy will be the Age when the Ladies may modestly pretend to Knowledge, and appear learned without Singularity and Affectation!'[1]

Euphrosyne and Cleonicus were fictional characters, invented in the mid-eighteenth century by the lecturer Benjamin Martin to make his books on natural philosophy entertaining as well as instructive. An expert at marketing and self-promotion, Martin aimed his teaching at the growing number of young people – women as well as men – who were fascinated by new scientific ideas, but lacked the mathematics or the money to study them academically. Educational writers like Martin were keen to enlarge their audience (and their profits) by addressing women. There was no need to spell out the sub-text: if even women can grasp

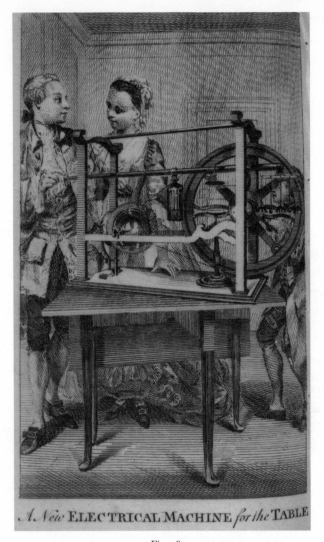

Fig. 18
Euphrosyne learns about electricity.
Benjamin Martin, *The Young Gentlemen's and Ladies Philosophy* (2 vols, London, 1759–63).

these scientific ideas, then any man should be able to understand them.'[2]

The clever brother instructs his adoring younger sister. Cleonicus and Euphrosyne were updated versions of a familiar couple, the learned philosopher and the naive young pupil – traditionally, of course, a boy – who feeds his teacher all the appropriate questions. One of Martin's rivals, a Fellow of the Royal Society called James Ferguson, taught astronomy to a wealthy young girl, an experience that inspired him to create another sister-brother pair, Neander and Eudosia. When Neander came back from Cambridge, Eudosia begged her brother to teach her astronomy, although she had a serious problem. 'But shall I not be laughed at,' she worried, 'for attempting to learn what men say is fit only for *men* to know?' Neander reassured her – and their readers – that any right-thinking man would encourage the many women who were eager to study science.[3]

Ferguson and Martin realised that there had long been many real-life sisters eager to study the new science, women like Anne Conway who begged her brother to keep her up-to-date with what he was learning at Cambridge. Conway was unusual because her work was published and her correspondence survives, but many other women must, like her, have borrowed their brothers' books.

Catherine Wright, a diplomat's wife from Plymouth, was one such sister who took advantage of her brother's privileges. His tutor, she reported, 'was a Man of Learning & fond of me, consequently gave me a taste for, & an Earnest desire after Studies not suitable to my Sex. What I learnt with him, Opend my Mind sufficiently to give me a pleasure in the Conversation of Men of Letters, & Books which I read at Stolen Hours.' Isolated in Devon after her marriage, she experimented on local minerals and shellfish, and embarked on a scientific correspondence course with William Withering, a famous doctor. Saddled with time-consuming family duties and an ailing husband, Wright repeatedly worried about her own insufficiency and the inappropriateness of her studies. She begged Withering not to 'expose my follies' or think her vain: she knew only too well that 'very few of our Sex Ever Attain to the Learning of a School Boy'.[4]

Withering's female correspondents resented being excluded from academic studies. Wright wrote tartly that 'the Generalty of men have Agreed that Women ought to be kept in perpetual Ignorance & the most profound Darkness'. Molly Knowles, wife of a fashionable London doctor, was still more outspoken: '*Women* to possess understandings of "masculine strength", is an idea intolerable to most men bred up amongst each other in the proud confines of a College.' Knowles sarcastically predicted a time when men would no longer monopolise learning: 'as general education increases, Scholars will more & more discover to the confusion of their pride, that genius is shower'd down on heads, as seemeth to Heaven good, whether drest in caps of gauze or velvet – in large grey wiggs, or small silk bonnets.'5

Another real-life Euphrosyne was Elizabeth Tollet, who – like Anne Conway – complained bitterly about being left behind with only her books for company while her brothers went to university. Like Émilie du Châtelet, Tollet benefited from having an enlightened father who ensured that she was skilled in Newtonian mathematics and astronomy, as well as being learned in literature and the classics. To this studious girl, it seemed unfair that her younger brother Cooke should be at Cambridge while she remained trapped inside the walls of the Tower of London, where her father lived and worked. Convinced that the 'gay and courtly' Cooke would not take proper advantage of the educational opportunities she craved for herself, Elizabeth compared their fates:

> All day I pensive sit, but not alone;
> And have the best Companions when I've none:
> I read *Tully's* Page, and wond'ring find
> The heav'nly Doctrine of th'Immortal Mind . . .
> Thrice happy you, in Learning's other Seat!6

Iceberg tips of evidence are all that remain, but these snippets suggest that, contrary to the conventional picture, many women shared Tollet's enjoyment of mathematics and natural philosophy during the seventeenth and eighteenth centuries. There are brief references to other individuals, such as the unnamed ten-year-old girl who astounded the Dublin Philosophical Society with her knowledge of 'Algebra, Mechanicks, the Theory of Musick, and Chronology or the Calendar'. As well as enduring interrogations on astronomy and geography, she 'was examin'd before ye Soc: in ye most difficult propositions of Euclid, w^{ch}, she demonstrated with wonderful readines'. When Polly Stevenson, the daughter of Benjamin Franklin's landlady, begged him to teach her science, he sent her a book similar to Martin's and, through letters, directed her studies for several years. Their correspondence covered a wide range of topics – barometers, tides, fireplace design. Then there are all the anonymous women who insisted that their favourite journal, *The Ladies' Diary*, provide mathematical puzzles for them to solve. In response to female demand, the editor published more and more calculations and enigmas, many of them contributed by the readers themselves.[7]

What became of all these young women? A few of them confided their frustration to their diaries and notebooks, leaving behind tangible evidence of their learning and their interest in science. Some explicitly chose a strategy that would enable them to continue their scientific activities – they decided to live with men who were engaged in experimental research. Many more probably abandoned their earlier enthusiasm as they became submerged in the daily round of caring for children and husbands. But because experiments were being carried out at home, and not in a distant laboratory, even women with no previous scientific education inevitably became involved. Some of them were co-opted as technical assistants, others as editors, classifiers, translators or illustrators. All of them had to negotiate ways of accommodating research projects within a domestic environment, traditionally a female realm. Although only fleetingly acknowledged (if at all), these hidden women were involved in family experiments and affected the pattern of science's history.

Concealed behind Euphrosyne's electrical machine (Figure 18) stands a diminutive servant turning the handle. Scarcely mentioned in the text, John is one of those normally invisible assistants who was essential for making the machine work. As well as all the servile Johns who have been forgotten, experimenters also relied on their female relatives for help, although they often omitted to credit them for their essential labour. A few women managed to make their existence visible. Their stories must stand in for all those others that have been lost.

Chapter 6

Jane Dee / John Dee

The whole duty of the wife is referred to two heads. The first is to acknowledge her inferiority: the next, to carry her selfe as inferiour . . . If euer thou purpose to be a good wife, and to liue comfortably, set downe this with thy selfe. Mine husband is my superiour, my better; *he hath authority and rule ouer mee: Nature hath giuen it him, hauing framed our bodies to tenderness, mens to more hardnesse. God hath giuen it him, saying to our first mother,* Euah, Thy desire shall be subject to him, and hee shall rule ouer thee . . . *Though my sinne hath made my place tedious, yet I will confesse the truth,* Mine husband is my superiour, my better.

William Whately, The Bride Bush, 1617

Euphrosyne (Figure 18) was brought up in a world lit by candles and oil-lamps, so that the rhythms of her daily life were governed by the sun rather than by clocks. As the servant John turned the handle of her brother's electrical machine, Euphrosyne was captivated by the mysterious lights that seemed to materialise out of thin air as if by magic. It is still exciting to sit like her in a darkened room and watch the globe of an electrical machine start to shimmer. For Euphrosyne, who had never before seen electrical sparks and glows, it must have been a quite extraordinary experience.

'On my word, *Cleonicus*,' she exclaimed, 'if you were to shew these Experiments in some Countries, with a black Rod in your Hand, and a three-corner'd Cap, and a rusty furred Gown on, they would certainly take you for a Conjurer, and believe you had the Art of dealing with the Devil, beyond even *Sydrophel* himself; for they could not possibly believe such Things were to be done by the Power of Nature, as you now shew by this small machine.'¹ Speaking through Euphrosyne, the teacher Benjamin Martin was hammering home the message that natural philosophers – especially English ones – relied on reason, not on magic, for their performances. Witchcraft and wizardry might, Euphrosyne marvelled, be good enough as explanations for ignorant people, but not for those illuminated by the electric light of Enlightenment rationality.

Euphrosyne's *Sydrophel* was a well-known fictional sorcerer, but many of her readers would immediately have thought of John Dee, an Elizabethan natural philosopher with a notorious reputation. Modern scholars present him as an eminent mathematician, but he was vilified by his enemies as 'Doctor Dee the great Conjurer'. This blackwashing process intensified during the following centuries, and in Euphrosyne's time, Dee was renowned as an outrageous charlatan of a bygone age. By exaggerating some aspects of Dee's life, propagandists for the power of reason cast him as an esoteric visionary, and so made their own arguments seem vastly preferable. Dee still symbolises the dark forces of arcane magic which have been swept away by scientific experiment and rationality. But magic sells well, and Dee's story can be romanticised by emphasising appealing ingredients, especially his conversations with angels . . .

John Dee was Queen Elizabeth's 'angel conjuror' who wore a black skullcap and 'put a hex on the Spanish armada which is why there was bad weather and England won'.² The early part of his life gave no clear indication of this extraordinary future. He read mathematics and astronomy at Cambridge, where he became one of the first Fellows of Trinity College; he also studied abroad, and lectured on Euclid and geometry at the Sorbonne in Paris. Dee even became a country clergyman – yet was this perhaps a cover for more arcane pursuits?

After turning down offers of university lectureships at Paris and Oxford, Dee embarked on an unorthodox way of earning his living – setting up a laboratory within his own home and relying on wealthy patrons for support. He openly rebelled against the traditional scholarship of the universities. Instead of studying Aristotle, Dee turned to mathematics and experiments and chose to investigate the natural world. Rather than remaining in the old-fashioned English university system,

Dee preferred to explore the exciting subjects being taught by Europe's leading learned men – astrology, numerology, alchemy and the Kabbalah. Mathematics, Dee learned from these magi, was a unique international language that operated on several levels. Most obviously, mathematics was a practical subject, vital for navigation, surveying and building. But more subtly, mathematics provided the key for entering the sphere of magic.

Dee's reputation as a magician overshadowed his more conventional achievements. When things went wrong, it was only too easy for his critics to accuse him of summoning up evil spirits. Dee's career depended on who sat on the throne. When Mary was queen, he was condemned for using his magic skills to plot against her and was made a prisoner at Hampton Court. In contrast, he flourished under Elizabeth I, who admired Dee enormously. Elizabeth employed him to work out the most propitious day for her coronation, and once she was queen, appointed him her court astrologer.

Under Elizabeth's protection, Dee's influence spread and his wealth grew. He converted his large, rambling house at Mortlake (about ten miles west of London) into the unofficial equivalent of a private research institute. Conveniently situated on the Thames near the royal palaces at Richmond and Hampton Court, Dee's home attracted many distinguished visitors. To support his scholarly studies, Dee built up England's largest private library, which was so splendid that Elizabeth rode over on horseback to admire it. (Many of his 4,000 books and 700 rare manuscripts are now in the British Museum.) As well as buying books and manuscripts, Dee poured money into instruments and hired assistants to help him in his research.

His tasks included reforming the calendar and giving navigational advice about voyages of exploration; he also became involved in some tricky political issues, such as deciding the legal basis for claiming Greenland and North America. Wealthy gentlemen paid Dee to calculate their astrological horoscopes, and he published learned Latin treatises on mathematics and astronomy. For practical tradesmen, he wrote in English, providing important instruction books on mathematics, navigation and geography. Dee's skilled knowledge of navigation made him a valuable asset as England expanded her overseas trade. He had good connections at court and was well regarded in aristocratic circles. His future seemed secure.

By his mid-fifties, Dee was England's most famous scholar, but he was dissatisfied. 'All my life time I had spent in learning,' he moaned. 'And I found (at length) that neither any man living, nor any Book I could yet meet withal, was able to teach me those truths I desired, and longed for.' Dee aspired not only to learn more about the world, but also to understand God. He prayed for help, and in 1581 he established contact with God's angels. To make his records of these spiritual communications more reliable, he decided to operate through a medium. After rejecting some obvious frauds, he found the perfect intermediary, a man called Edward Kelly who claimed to have special powers of communication. At frequent intervals over the next eight years, Kelly gazed intently into a crystal globe and reported hearing hundreds of angelic pronouncements. Totally confident of Kelly's authenticity, Dee faithfully recorded their conversations – he had successfully communicated with spirit beings from another plane of existence.[3]

After Kelly and the angels moved in, life in the Mortlake household was no longer the same. Long days were devoted to transcribing declarations from Raphael, Uriel and other holy messengers. Dee persevered, convinced that he was truly gaining access to the beneficial knowledge that he sought. However, the angels became less cooperative, perhaps because Kelly was worried about being arrested for an earlier crime. As tensions escalated, Dee was rescued by one of his many admiring visitors, a wealthy Polish nobleman, who invited the combined Dee and Kelly families to visit him in Poland. Unfortunately, the funding that he provided for their alchemical research soon ran out, and they were forced to go on the road displaying their skills.

Sick, old and short of money, Dee eventually became disillusioned by Kelly's underhand activities and abandoned him in Prague. He returned to England, where Elizabeth protected him while she was still alive. Dee eventually died at Mortlake in 1608, by then so poor that he resorted to selling off his library books for food.

Angels and magic, navigation and astrology, success and disillusion: a sad and romantic story, but surely nothing to do with science? The answer depends on what sort of history you want to tell. One route into the foreign country of the past runs along well-known roads, singling out famous scholars like Isaac Newton or Robert Boyle who led comfortable lives, moved in prestigious academic circles and published books that seem to be direct precursors of modern science. But other tracks exist as well, which lead to men who are less well-known and who were involved in activities not now regarded as scientific. Dee railed against the smear campaign launched by his enemies. Why, he demanded, should I, an 'honest Student, and Modest Christian Philosopher', be vilified 'as a Companion of the Helhoundes, a Caller, and Conjurer of wicked and damned Spirites?' He made this bitter complaint in a learned disquisition on the mathematical sciences – hardly the publication one would expect from a wizardly magician. Modern historians have rewritten Dee's story to include him within the history of modern science.[4]

Dee was an expert on navigation, and the practical mathematical information in his books was just as important for later sciences (such as geomagnetism and meteorology) as the theoretical debates of scholarly, sedentary natural philosophers. And the complicated astrological calculations he carried out for his clients were far closer to mathematical astronomy than to a newspaper columnist's glib horoscope. At the time, astrology and certain kinds of magic were highly respected. It is unfair of us to sneer at them because they stemmed from a view of the universe that has now been rejected.

Although their activities were often misinterpreted, men like Dee called themselves natural magicians because they wanted to make it clear that they were not dealing in black magic, which was the art of communicating with supernatural demons and witches. Instead, they operated by tapping in to astral influences and they set themselves the highest aspirations – wisdom, virtue, closeness to God. According to their model of the universe, the spiritual and the material realms were intimately bonded together, so it made sense to seek guidance from angels. Especially in continental Europe, magic and the occult were legitimate topics to study, although there was a wide range of different approaches. With centuries of scholarship behind them, Renaissance magi constantly experimented to test how different substances and techniques could help them achieve the effects they sought. Not so differently from modern scientists, they were trying to harness nature's hidden powers.

Dee has been vilified as an evil magician, yet he was Elizabethan England's most eminent mathematical astrologer whose interest in spirits was shared by leading scholars. Even Newton gained many of his insights from poring over alchemical treatises which – like the books Dee studied and wrote – portrayed a harmonious, interconnected universe. Dee published important texts on mathematics, navigation, geography, subjects that directly influenced the new theories and instruments developed in the early Royal Society. As Sprat's frontispiece (see Figure 8) implies, Fellows of the Royal Society later adapted practical implements of the type described by Dee and relabelled them as instruments of natural philosophy. Similarly, although alchemy and astrology have now been discredited, chemists and astronomers built on their traditional techniques that had been developed over centuries.

In addition to this intellectual significance, there are other reasons for reinstating Dee in science's history. For one thing, unlike many of his contemporaries, he was a natural philosopher who carried out experiments as well as reading books. Furthermore, examining the origins of modern science means knowing not only what people were studying, but where they were doing it. Dee is important because he worked at home rather than in a monastery or a university.

Four hundred years ago, in Dee's lifetime, there were no university courses on science, no large research laboratories, not even any scientific societies or institutions. In a sense, men like Dee were engaged in two types of experimental programme. Most obviously, they were using their new instruments to explore the natural world instead of relying on the authority of ancient texts such as Aristotle and the Bible. But in retrospect, we can see that they were also embarked on a second kind of experiment. By using their own homes as the place to pursue their investigations and earn some money, they were trying out a new style of living. Searching for scientific knowledge involved changing traditional social patterns. Understanding how science started entails examining how families coped with the demands of natural philosophy as the structure of society altered.

By the middle of the seventeenth century, there were several scientific partnerships in private houses, such as Robert Boyle and his sister Katherine Jones, John Evelyn and his wife Mary Browne, and Robert Hooke and his niece Grace Hooke.[5] One way of thinking about these scientific households is to regard them as social experiments, attempts to forge a new type of small community. These trials in living had momentous outcomes. In some respects they failed, because by the end of the nineteenth century most scientific research was taking place outside private settings. Although some scientists – including Charles Darwin – were still carrying out research in their own family homes, men (and by then, a few women) had generally moved out to work in universities, industrial laboratories and scientific academies. But in the long term, these experiments were hugely successful, because natural philosophers flourished and found ways of making their investigations financially profitable. Initiatives like Dee's provided the foundations of our modern technological society, in which professional science and financial interests are bonded together.

Inevitably, when science was happening at home, women were involved. Compared with women like Katherine Jones, Mary Browne and Grace Hooke, Jane Dee does not seem herself to have been involved in experimental research. Like many other wives, her existence is generally ignored. Yet John Dee's livelihood, the new type of scientific career he was forging, depended on her cooperation. The integration of science within society, and the upheavals in domestic life that this entailed, only became possible because of these silent partners. So another way of telling John Dee's story is to place more emphasis on his wife Jane.

But first, a comment on names. One way of restoring forgotten women is to dignify them with their own surname instead of their husband's. All very well and politically correct, but there are objections to this strategy. For one thing, it is more historically sensitive to describe people as they and their contemporaries knew them, and this almost inevitably means referring to wives by their married name. However objectionable it may be to modern feminists, these women did derive part of their identity – psychologically as well as legally – from being married to a particular man. Reverting to their maiden name (which in any case, a woman inherited from another man, her father) would conceal a status that was important to them. There is also a practical problem of communication: does it really make sense to confuse readers by referring to Mary Godwin rather than Mary Shelley, or to Marie Skłowodska rather than Marie Curie? There are no easy answers to these problems. However, what does seem vital is to treat men and women symmetrically, and not condescendingly use first names only for women. So, at the risk of being cumbersome, in this book wives and husbands are generally called by both their names.

Virtually none of Jane Dee's own words survive, but her experiences can be partially reconstructed by interpreting existing documents in new ways.

Jane Dee (née Fromond) (1555–1605) deeply resented Kelly and the angels. No doubt she often pointed out that Kelly had appeared bearing a false name but minus his ears, which had been cut off as a punishment for forgery. A couple of months after Kelly arrived, John Dee made a note (later scored through) that his wife was 'in a marvellous rage at 8 of the clock at night till $11\frac{1}{2}$ and all that night and next morning till 8 of the clock, melancholic and choleric terribly'.[6] Ironically, it is only because of John Dee's faith in the angelic exchanges mediated through Kelly that so much information about her feelings and activities survives. For thirty years, John Dee kept an intimate diary, in which he aimed to explore systematically how changes in the heavens affect life on Earth. Intermingled with dictated diagrams and speeches received through Kelly are details of births and journeys, illnesses and debts, visitors and quarrels – unique evidence about the daily life of a philosophical wife.

5 February 1578: 'My marriage to Jane Fromonds'. A terse diary entry. Other comments hint at hidden resentments on both sides. When he became a father for the first time on his fifty-second birthday, John Dee recorded the astrological conditions of the conception and the birth. Similarly, in the interests of research, he observed her illnesses, meticulously noting any cramps or the colour of her vomit. According to him, she was often 'testy and fretting', impatient, or angry with the maids – but if only we could hear her side of the case, she could surely give reasons for her fits of temper. Many partnerships would have dissolved under such intense proximity, but Elizabethans took church marriage vows very seriously. Despite their rows, the couple stayed together until her death almost thirty years later, and their marriage generated a total of eight children ('generated' is a carefully chosen word: some of the babies died, and two of them were associated with a tumultuous period when the angels ordered John Dee and Kelly to swap wives).

Jane Dee, a gentleman's daughter from the court of Elizabeth I, found herself in a strange position. Neither she nor her husband fitted into any existing category, and there were no established guidelines for how either of them should conduct themselves. They got married over eighty years before the foundation of the Royal Society, and no collective identity had yet been established for natural philosophers. Only later did this become a recognised term embracing a large variety of men who used experiments and instruments to study nature, often with the overarching goal of learning more about God. Moreover, at this time, most scholars were single. In England, many of them were secluded in monasteries or universities, neither of which allowed their inhabitants to marry. In contrast, European magi did have private accommodation, but they avoided marriage in order to keep their souls pure. John Dee broke away from all these conventions by living at home, marrying and trying to support his family from his scientific work. Both Jane and John Dee were involved in forging this new type of married existence, which later became very common amongst scientific researchers.

Jane Dee was torn between conflicting codes of behaviour. As a virtuous wife, she should submit to her husband's wishes, especially as he was nearly thirty years older than her. For women, marriage often meant being transferred from a father's authority to a husband's. On the other hand, gentlemen were usually occupied at court or in outdoor activities, leaving their wives in charge of the household (Figure 19). Jane and John Dee were forced to negotiate the ground rules for a new kind of domestic experimental partnership, one that would become increasingly common during the seventeenth and eighteenth centuries.

Elizabethan men prescribed separate roles for themselves and their wives: 'Whatsoever is done without the house,' intoned a guidebook to marriage (written by a man), 'that belongs to the man, and the woman [is] to study for things within to be done.' Confined at home while their husbands went out to work, women were forced into economic dependency, yet did exert authority inside their houses. Figure 19 shows a virtuous housewife who, resembling a less privileged version of Elizabeth I, is reigning over her domestic realm. The dog at her feet symbolises marital fidelity, while Father Time is preparing to crown her with a laurel wreath as a reward for good behaviour.[7]

As in many similar pictures and pronouncements, there is a certain

Fig. 19
An idealised Elizabethan housewife.
A good housewife, c. 1600, hand-coloured woodcut, *c.* 1750.

amount of wishful thinking going on. Apprehensive about their women's strength, Elizabethan men proclaimed their own power. The verses here include a facetious couplet pointing to women's inherent Eve-like nature:

> Such Wives as this I doubt not but there are;
> But like the black Swan they are wondrous rare.

This picture illustrates an idealised version of Elizabethan domesticity. Powerfully placed at the centre of her tidy, well-organised house, the woman is spinning wool, the cliché for feminine work. But she was also responsible for superintending the servants – hence the girl sweeping in the background – and for minding her children, here neatly divided into the appropriate gender skills: embroidery for the two girls, and book learning for the boy. Women also took charge of the family's health, and exchanged recipes for medicines and tonics. When a baby was born, a network of women rallied protectively round the new mother for several weeks, creating a nurturing female environment from which men were excluded.[8]

Even Elizabeth I had been taught to cook and sew as a girl, but when she reached the throne she devised strategies for operating as a woman in a male world. Jane Dee, an exile from Elizabeth's court, had to find her own ways of resolving the conflicts between ruling over her household and accommodating her husband's research activities. The lines of authority were blurred. Inside their home, who was responsible for the experimental workers: the wife or the husband? Who was meant to cope when, as happened on several occasions, John Dee's servant George fell off his ladder? The children were another source of contention. Unlike other Elizabethan gentlemen, John Dee was no absent father figure, but should he be involved in looking after them? He struggled to reconcile being the sole breadwinner with staying at home to work instead of going out. He must often have wondered whether he should choose a more orthodox, reliable way of supporting his family, one that would not involve constant worries about money and security. No wonder that Jane Dee scared her husband by flying into 'mervaylous rages'. She was encountering problems that became more and more frequent as entrepreneurial philosophers converted their houses into schools, workshops and research centres.

*   *   *

As well as managing the domestic staff – eight of them at one stage – Jane Dee had to look after her husband's live-in assistants and apprentices, supervising their food, health and accommodation. She must have been relieved that some of them experimented in the outhouses, which had been converted into alchemical workshops. However, Kelly was based in John Dee's study, right in the middle of the house, as though he were an invader into her territory. Like Kelly, some of these assistants had dubious pasts, and John Dee deliberately picked moody men whose melancholic temperaments would make them more sensitive to astral influences. Sometimes they were accused of crimes, and magistrates visited the house to take them into custody. The domestic servants could also be troublesome: a couple of maids twice set their room on fire, and the children's nurse explained away her erratic behaviour by claiming that she had been tempted by an evil spirit. Jane Dee found it difficult to discipline such a mixed household, and sometimes took out her rage on her own children – once she hit her eight-year-old daughter so hard that her face bled.

Money was a major issue. Sometimes Jane Dee pulled strings by going back to the court and pleading for help from Elizabeth I. The couple often settled wages and paid bills in arrears. John Dee had no regular income, yet Kelly threatened to leave without frequent salary rises, while the servants, staff and wet nurses were further steady drains on the family economy. John Dee kept buying expensive books and equipment and insisted that his wife lay on generous hospitality for the stream of important visitors. Jane Dee was in a Catch-22 situation: she had to convince potential patrons that her husband deserved sponsorship, yet at the same time conceal her limited budget by demonstrating that together she and her husband ran a flourishing, efficient household.

One time, in desperation, she overcame her loathing of Kelly to remind the angels (through him) of their poverty. Surely, she asked, she should not have to pawn their possessions in order to finance philosophical experiments? Kelly summoned up a handsome young man in a white coronet, who tartly reminded her to be faithful and obedient, and recommended that she go back to her sweeping. 'Give ear unto me, thou woman,' the spirit ordered her through Kelly, 'is it not written that women come not into the synagogue?'9 The synagogue? Kelly and John Dee carried out their experiments behind double closed doors, making the study an inner sanctum reserved for men; even the outbuildings were the preserve of a gloomy assistant who operated the alchemical stills. Instead of ruling over her own house like other Elizabethan women, Jane Dee was being squeezed into restricted quarters. She had to organise her rooms not only to make domestic operations run smoothly, but also to give John Dee and Kelly privacy for their conversations with the angels.

Life must have been exceptionally chaotic when the entire household – around twenty assorted Dees, Kellys, children, servants and assistants – trailed round central Europe. Even at Mortlake, it had been hard to coordinate conflicting schedules and make John Dee's experimental investigations fit in with the other family activities. During one exceptionally long session with the archangel Michael, John Dee 'axed if I should not cease now, by reason of the folk tarrying for supper'. The scene is easy to envisage: hungry children, an angry wife, the food drying out . . . how many other wives have been left to entertain the children while their husbands finished off an experiment? But Kelly was adamant. Speaking through him, Michael instructed John Dee to 'lay away the world, continue your work' – the meal had to wait.[10] Like the *Good housewife* of Figure 19, Jane Dee was trying to impose regularity in her own household, yet her routine domestic timetables were being overthrown by the demands of research.

Sex was another source of conflict. As a Renaissance magus, John Dee should have kept himself pure by abstaining, especially if he wanted to establish clear lines of communication with the angels. Sometimes he noted that she had initiated sex; was this perhaps to absolve himself from the responsibility of tainting his scholarly soul? On the other hand, Jane Dee knew that her obligation as an Elizabethan wife was to bear him many children and so help atone for Eve's original sin by suffering the agonies of childbirth. But she was also required to be obedient to her husband, and sometimes his desires conflicted with her duty – especially when the angels, communicating their instructions via Kelly, insisted that John Dee and Kelly exchange wives. (One interpretation of this unusual order is that Kelly was desperate for children, but his wife was infertile.) Jane Dee was horrified. In a stormy session, she 'fell a weeping and trembling', her husband recorded, but 'I pacified her as well as I could'. John Dee apparently did try to talk the angels out of the plan, but eventually resorted to persuading his wife that the virtuous path lay in accepting God's command. Reluctantly she accepted, although she did insist that they all four share the same room so that her husband would be nearby.[11]

Jane Dee coped with the incompatibility of marital obedience and fidelity by praying 'that God will turn me into a stone before he would suffer me, in my obedience, to receive any shame'. Her husband resolved the conflicts between natural philosophy and Elizabethan conventions by converting Jane Dee into the object of his astrological research. By recording her body's behaviour, he could study her fertility and collect data for improving his horoscope techniques. He systematically recorded her periods, moods and illnesses; to indicate when they had intercourse, he devised a special symbol to represent the combination of Venus and Mars. Natural philosophers often experimented upon themselves, and many others must also have kept intimate diaries about their wives, trying to impose orderly patterns on female bodies (the relationships between menstruation, fertility and conception were unresolved at this time). Charles Darwin's sons, for instance, kept a register of their own sexual experiences because they were interested in exploring eugenic methods of 'improving' the population.

When the wife-swapping agreement with Kelly came into effect, John Dee coolly noted the event in rhyming Latin: *Pactum factum* (the agreement has been carried out). But someone (Jane Dee, perhaps?) has cut out the rest of that page in his diary. Nine months later, John Dee recorded that 'Theodorus Trebonianus Dee was born, with [Mercury] ascending in his horoscope.' Although Theodore was presumably Kelly's child, John Dee claimed paternity, perhaps because he felt that this youngest son represented the culminating achievement of the angelic seances.[12]

Soon after Kelly arrived at Mortlake, he dreamed – or was it the angels speaking to him? – that Jane Dee was killed when she fell over a gallery rail. Despite this evidence of Kelly's hostility, she survived for more than twenty years, dying in Manchester a few years before her much older husband, probably from the plague. Several of her children may well have died with her, although this can only be inferred from the absence of their names in John Dee's diary. He did, however, record the death of the angelically conceived Theodore at the age of thirteen.

*   *   *

Even by Elizabethan standards, Jane Dee's life was extraordinary. Few wives had to tolerate a host of angels who excluded her from parts of her own home, dominated meal times and forced her into her rival's bed. Yet instead of focusing on these exotic incidents, it is more rewarding to think about other aspects of the Dee ménage. John Dee's diary shows his voyeuristic control over his wife's body, yet it also reveals the daily experiences that would become increasingly common during the following two centuries, as scientific research moved inside women's own homes.

John Dee's interests seem bizarre to us because we have rejected Renaissance visions of a universe bound together with occult virtues. Many learned Elizabethans accepted the value of astrology and alchemy, but such knowledge became excluded from legitimate science. From our perspective, John Dee seems to be a late representative of vanished systems of belief. However, he could also be seen as a social pioneer, an entrepreneur who strove to integrate his philosophical activities with his domestic life. Viewed in that way, John Dee appears as an early example of the natural philosophers who helped to make science into a collaborative, public enterprise.

Renaissance scholars worked in private studies and only a select group of privileged men had access to their arcane knowledge. In contrast, modern scientists receive large sums of money to carry out research and publicise their results. Taking experimental philosophy into the public arena was a key move in this transformation. In the early eighteenth century, Mr Spectator – the essayist Joseph Addison – wrote that 'I shall be ambitious to have it said of me, that I have brought Philosophy out of Closets and Libraries, Schools and Colleges, to dwell in Clubs and Assemblies, at Tea-Tables, and in Coffee-Houses.'[13] He was continuing an initiative that had been launched more than a hundred years earlier by John Dee and his contemporaries.

Bringing experimental research into private homes meant that women became involved. Details about Jane Dee's life have survived because her husband kept an unusually detailed diary, but many of her experiences illustrate the difficulties that other women must have encountered. Scientific wives often had to look after residential assistants and students, providing their food and coping with their misdemeanours. How many of them were, like Jane Dee, left to placate the children and keep the dinner warm while their husbands finished off a piece of work? How many of them watched the family finances being poured into new instruments, books and chemical supplies? How many of them, as Benjamin Franklin wrote, 'threatend more than once to burn all my Books and Rattling-Traps (as she calls my Instruments) if I do not make some profitable Use of Them'?[14]

Every woman's life is unique, and it is hardly likely that many families were governed by directions from angels. However, during the following three centuries, other wives, sisters and daughters invented their own ways of reconciling the conflicting demands of domestic duties and scientific investigations. If they had not done so, science as we know it would not exist and the technological world we live in would be very different.

Chapter 7

Elisabetha Hevelius/Johannes Hevelius

Neander: *Why do you sigh,* Eudosia?

Eudosia: *Because there is not a university for ladies as well as for gentlemen . . .*

Neander: *You are far from being singular in this respect. I have the pleasure of being acquainted with many ladies who think as you do. But if fathers would do justice to their daughters, brothers to their sisters, and husbands to their wives, then there would be no occasion for an university for the ladies . . . the ladies would have a rational way of spending their time at home, and would have no taste for the too common and expensive ways of murdering it, by going abroad to card-tables, balls, and plays: and then, how much better wives, mothers, and mistresses they would be, is obvious to the common sense of mankind.*

James Ferguson,
The young gentleman and lady's astronomy, 1768

Far away on the planet Venus an insignificant crater commemorates one of those forgotten women of the seventeenth century – scientific wives. In the past, stars and comets were often named after monarchs, but now the heavens have become more democratic. A crater thirty miles wide on Venus: this small and distant feature pays tribute to Elisabetha Hevelius (1647–93), a Polish astronomer whose observational expertise was known throughout Europe at the end of the seventeenth century.

Even as a child, Catherina Koopman (her maiden name) had wanted to study the stars, and when she was sixteen she realised how to achieve her aim. Johannes Hevelius, Poland's most famous astronomer, must have seemed very elderly to her, but his wife had recently died – the ideal opportunity for an ambitious teenager. According to the romanticised version of their courtship, she wriggled her way into his affections. Reminding him of a childhood promise he had made to teach her astronomy, she claimed to be overawed by the stars and his wisdom. Within a few months they were married, and she took on a new name and a new identity: Catherina Koopman became Elisabetha Hevelius (his first wife had also been called Catherina, but no record remains of who instigated the switch to Elisabetha). Together they observed the heavens for more than twenty years (Figure 20). 'An old peevish gentleman' one visitor called Johannes Hevelius.[1] Perhaps Elisabetha Hevelius also found him trying, but in compensation she could indulge her fascination with science, meet some of the world's leading astronomical experts, and be confident that she was internationally recognised as her husband's skilled collaborator.

After Johannes Hevelius's death, Elisabetha Hevelius organised the publication of an important book, a giant celestial atlas that mapped 1,564 stars and included seven new constellations. He had diplomatically named one of these constellations the 'Shield of Sobieski' after the Polish royal family, an astute move to ensure that a common unspoken bargain came into operation: in exchange for financial support, astronomers would immortalise their patrons. Perhaps Johannes Hevelius had been following the example of Galileo, who converted the satellites of Jupiter into Medicean stars as he negotiated his career at the Medici court in Florence.

When she became a widow, Elisabetha Hevelius needed to ensure that the Polish king continued paying her a pension. She too demonstrated her skill in the patronage game. Making sure that the title of her atlas mentioned 'the Sobieskan Firmament', she drew up a long ingratiating preface. First came the introduction – seven lines, packed with superlatives, to salute the King – and then further on, a not-so-discreet reminder that in her husband's heavenly map 'glows the family shield of your Majesty marked with the victorious cross and the seven stars'. In addition, like many other scientific writers of this period, she included a

Fig. 20
Elisabetha and Johannes Hevelius observing with their great sextant.
Johannes Hevelius, *Machina Cœlestis* (Danzig, 1673).

symbolic frontispiece (Figure 21) to celebrate her royal patron and cement
their alliance.[2]

Although Elisabetha Hevelius had organised the book's publication, she
made it clear that she was merely standing in for her dead husband. Amidst
her flowery platitudes hymning the King's military victories, she remarked
that it was appropriate for her sex 'to admire things belonging to sublime
altitudes, but not to meddle with them'.[3] Similarly, the allegorical frontispiece

132

Fig. 21
Johannes Hevelius pays tribute to Urania and his
astronomical predecessors.
Frontispiece of Johannes Hevelius, *Prodromus Astronomiæ
& Firmamentum Sobiescianium* (Danzig, 1690).

conceals Elisabetha Hevelius's contributions by portraying a conventional
view of who mattered in astronomy . . .

Seated on her throne, Urania – the goddess of astronomy – is flanked
by a heavenly throng of distinguished astronomers, including Copernicus,
Ptolemy and Tycho Brahe. Overhead, small cherubs carry banners
proclaiming the glory of God. Johannes Hevelius has apparently trudged
skywards from Danzig (now Gdansk), his home town, the wealthy trading
port that is lovingly depicted in the bottom right-hand corner. Bowing
down before Urania, he presents her with his star catalogue, propped up
on a cushion of clouds; in his right hand he clutches a real Sobieski
shield, another overt reference to the constellation he had christened.

Crowned with a glittering star, Urania clasps the sun and the moon as if she controlled them.

In Johannes Hevelius's left hand, the ornate brass instrument – a miniature version of the great sextant in Figure 20 – emphasises how his reputation rested on his exceptionally keen vision and powerful apparatus. He was 'the Prussian Lynx', the ruler of 'Startown' (*Stellæburgum*), the huge astronomical observatory that he had built on the roof of his own home. To reduce the optical distortions that plagued Galileo and earlier observers, Johannes Hevelius built extremely long telescopes and devised elaborate systems of pulleys to manipulate them and stop them blowing about in the wind. Priding himself (justifiably) on these technical innovations, he refused to adopt the new telescopic sights that were invented in England; nevertheless, he achieved superb results.

More than forty years earlier, in the same year that Catherina Koopman was born, Johannes Hevelius had published a treatise on the moon whose detailed maps remained the best available for a century. This book revolutionised astronomy by including elaborate, precise diagrams of the instruments that he had invented, together with full instructions for making them. Further major books followed – a study of comets, and a two-part comprehensive study of astronomy, which included an astronomical history as well as descriptions of his observatory and his observations.

Johannes Hevelius was an innovator twice over. His large, finely crafted instruments enabled him to observe lunar craters and mountains that had never been seen before; and he insisted on sharing his expertise, so that other astronomers could replicate his results by building exact copies of his apparatus. Himself a skilled craftsman, Hevelius supervised and trained his assistants in his own workshop. He prided himself on grinding his own lenses, explaining how he picked the finest glass and designed telescope cases that were light yet durable. By the time he died, 'the sharp-seeing Hevelius' was celebrated all over Europe for his accurate measurements.[4]

Urania occupies a prominent position in Figure 21, but her female presence is iconic. There is no reference to Elisabetha Hevelius, the woman astronomer who, like her husband's other collaborators and assistants, remains hidden behind the scenes. Johannes Hevelius pays tribute only to his famous male predecessors, the men who surround Urania and whose work (again impossible without the help of those invisible technicians) has made his own observations possible.

Johannes Hevelius hired professional iconographers to produce the allegorical frontispieces and decorations in his books, but his astronomical pictures were completely different. Unlike other astronomers, Hevelius insisted on engraving his own diagrams. He dedicated himself to making accurate images not only of the skies, but also of his instruments, his workshop and his observatory. He even employed his own printer to make sure that his instructions were closely followed. Instead of arguing with words, Hevelius relied on visual language for presenting his scientific arguments. He drew meticulous pictures that enabled his readers to share his own view of the heavens as if they were standing at his side.

Johannes Hevelius's unprecedented concern with realism means that we can be confident in his illustrations of his rooftop observatory. Although Elisabetha Hevelius's clothes in Figure 20 might look suspiciously glamorous, there is no doubt that she did actually use the instruments as shown. When Johannes Hevelius portrayed himself measuring stars in partnership with his wife, he was inviting their fellow astronomers to witness their daily routine – or to be more accurate, their nightly routine. Astronomical observation did, of course, take place during the hours of darkness: these pictures only appear to have been drawn during the daytime.

Adept at self-publicity, Johannes Hevelius presented handsomely bound copies of his books to distinguished people all over Europe. So unlike other scientific wives, Elisabetha Hevelius's astronomical activities were no secret because she could be seen in action. Figure 20 is one of three similar pictures, all of them unusual and powerful demonstrations of a woman's active involvement in astronomical observation. In contrast with the allegorical frontispiece of Figure 21, the real-life Elisabetha Hevelius was no symbolic Urania, but a practising astronomer.

Only a few bare facts survive about Elisabetha Hevelius's personal life: she was the daughter of a Danzig merchant, she bore four children, and – thanks to her husband's devoted care – survived an attack of smallpox that left her badly scarred. For a woman, she must have been well educated, since Johannes Hevelius commented on her mathematical skills and she wrote letters in Latin, the international scholarly language.

However, Johannes Hevelius's pictures and data records provide unique evidence of her astronomical activities. By picking Johannes Hevelius, Catherina Koopman had astutely married a distinguished citizen, a brewer wealthy enough to build his private observatory across the roofs of three adjoining houses. Within a couple of years, she was spending nights up on the roof making astronomical observations, and she became his most trusted collaborator. Visitors from all over Europe came to admire the couple's sextant, telescope and other instruments perched up above the city near the banks of the river Vistula. Many of them met Elisabetha Hevelius, who regularly observed at her husband's side and entertained their guests.

Because she has left no personal documents, her own feelings about her status can only be inferred. Almost certainly she felt frustrated at being denied a university education, and so had taken the best option available to her – marrying a distinguished, if elderly, astronomer. Elisabetha Hevelius must have felt torn between managing her household, caring for her three surviving daughters and passing the hours of darkness up on the roof mapping the stars. Johannes Hevelius's first wife had taken care of the family brewery, and Elisabetha Hevelius probably had that responsibility as well. Included in the perks that came with the royal pension were a life-permit to sell beer and exoneration from paying dues to the Brewers' Guild; scientific tourists in Gdansk can still buy bottles of Hevelius beer with his picture on the label.

Johannes Hevelius's major instrument was the sextant shown in Figure 20, a large brass instrument used to measure the angle between two stars. Its cumbersome construction meant that two people were needed to operate it: even the handle that Hevelius holds is decorated with a miniature man and woman. Here Elisabetha Hevelius's task is to keep her end aligned on one particular star while her husband moves the adjustable radius towards a second one; the angle separating the stars can then be read from the curved scale. Making matters more complicated, the sextant can also be tilted by the system of ropes and pulleys. Here, the two observers are making final adjustments with fine screws.

Unlike many of his contemporaries, Johannes Hevelius did at least acknowledge the existence of his assistants. Although he rarely named them, he made two exceptions: his printer and Elisabetha Hevelius. He concealed his hired printer's true identity by referring to him as Typographus, but – in contrast – he singled out Elisabetha Hevelius's abilities. 'My dearest wife', as Johannes Hevelius called her in his published account of the sextant, was his diligent collaborator and a keen student of mathematics. In his opinion, he continued in an unusual declaration of equality, 'Women are definitely just as well suited to observing as Men.'[5]

In 1674, eleven years after the Heveliuses' marriage, John Flamsteed, Britain's Astronomer Royal, received an ingratiating letter from a young Oxford undergraduate who signed himself 'Your and Urania's most humble Servant tho' unknown, Edm. Halley'. Fifty years later, Edmond Halley was himself a world-famous astronomer who had inherited Flamsteed's position, but this was their first contact. Flamsteed and other English astronomers were introducing new sighting devices, which they claimed made their telescopes more accurate. Since Johannes Hevelius was renowned for his observations, someone was needed to vet the competition and compare the rival instruments. It may well have been at Flamsteed's instigation that Halley invited himself to Hevelius's Startown in Danzig.

Measurements started the very night that Halley arrived, and continued for almost two months. Johannes Hevelius regarded Halley's visit as the climax of his life's work, his opportunity to demonstrate once and for all that his methods were the best in Europe. For this international contest between scientific technologies, he wanted to use his most reliable observers – so he chose Elisabetha Hevelius. The records show that different combinations of couples took turns working with the sextant. Sometimes Elisabetha Hevelius collaborated with Johannes Hevelius, sometimes with the printer. She also partnered Halley, presumably teaching him how to use their instruments correctly. By the time that he left, Halley was convinced that the Hevelius couple could obtain results to match the English ones for accuracy.[6]

Elisabetha Hevelius was responsible for welcoming visiting astronomers to Startown, and she was probably glad to meet Halley, who was only nine years younger than her. Certainly they became friendly, since she commissioned him to order some fashionable silk clothes for her in London. Figure 20 and the other illustrations suggest that she took great care with her appearance. Even if silk dresses were not the most sensible clothes for chilly night-time work on a roof at the edge of the Baltic, she was obliged to wear fine gowns for dinners. Halley does not emerge well from the transactions over this purchase. After Johannes Hevelius died, Halley got very worried about being paid. Carefully converting his expenses into the right currency, he craftily suggested that Elisabetha Hevelius should send him some of her husband's books, a ruse that left him owing money to her.

Gossip circulated about Elisabetha Hevelius's night-time observation sessions with Halley. Why, people wondered, had Halley lingered far longer than anticipated in Danzig, apparently too busy even to send letters back to England? Years later, Halley's enemies were still insinuating that he had made Johannes Hevelius 'a Cuckold, by lying with his wife when he was at Dantzick, the said Hevelius having a very pretty Woman to his Wife, who had a very great Kindness for Mr Halley and was (it seemed) observed often to be familiar with him.'[7] Inevitably, historians have poured energy both into attacking and into defending her honour, scouring the sources to prove what is probably unprovable. Then as now, unconventional women posed threats to existing social hierarchies. The rumours may or may not have been true, but whether or not two people had an affair 300 years ago seems less interesting than reflecting on the ferocity of the debates. The original scandal highlights how Elisabetha Hevelius was under scrutiny as she negotiated appropriate ways to behave in a role that had few established codes. And the controversy's endurance indicates how other women were subsequently also obliged to show that respectability could be compatible with scientific research.

Without Elisabetha Hevelius, it is extremely unlikely that her husband's important celestial atlas would have been published. Perhaps, like Johannes Hevelius's voluminous correspondence, the hand-written manuscripts would have been sold off by their son-in-law for a pittance. However, it is unclear exactly how much work remained to be done when he died. Elisabetha Hevelius wrote a long letter in Latin asking London's Royal Society for help, but although her husband had been a Fellow for twenty-five years, this plea from a foreign woman was totally ignored. The Polish king was more forthcoming, and after some final tidying up, the observations and maps were printed – three books usually bound together in a single volume, the *Prodromus astronomiæ* (*Forerunner of Astronomy*), which included the most comprehensive and accurate star catalogue then available.

In her dedication to the king, Elisabetha Hevelius wrote that the book was the product of Hevelius's genius. It did, however, also contain a permanent record of the many years that she had herself devoted to astronomy, although this time there were no pictures of her. Like other domestic female astronomers, Elisabetha Hevelius's activities were now concealed from public view. She was absent from the allegorical frontispiece, and her presence disappeared inside Hevelius's book, which presented the observations as if they were his own. Even her body vanished – after she died, she was buried inside her husband's tomb.

The pictures of Elisabetha Hevelius in action with the sextant and other instruments on the rooftop (Figure 20) are fascinating not only because of what they reveal about her, but also because they vicariously illustrate other female astronomers who have disappeared with no such published pictures of their achievements.[8]

In England, Flamsteed's wife Margaret admired the older woman whom she had never met, but whose conduct she emulated. Like Elisabetha Hevelius, she observed the skies with her husband, carried out calculations, entertained important visitors and was also left as a young astronomical widow struggling to publish her husband's celestial atlas.[9] It seems likely that other English astronomers' wives became invisible assistants, although there are no well-documented examples of women in astronomical families before Caroline Herschel came to England towards the end of the eighteenth century.

In Germany, astronomical women were far more visible. Between 1650 and 1720, an astonishing 14 per cent of German astronomers were women, many of them collaborating at home with members of their family. When the Swedish astronomer Anders Celsius (who is now most famous for his temperature scale) travelled south to the German states in the early eighteenth century, he commented: 'I begin to believe that it is the destiny of all the astronomers with whom I have had the honour of becoming acquainted during my journey to have learned sisters.'[10]

The lives of some of these women mirror Elisabetha Hevelius's. Maria Eimmart, for instance, trained in her own father's observatory at Nuremburg, where she studied alongside his male apprentices. A skilled artist, she specialised in producing fine, detailed drawings of the sun and the moon. To secure her astronomical career, she married the physics teacher who had been appointed to manage her father's observatory. Her new husband was also presumably delighted with the arrangement, since when Eimmart's father died, he automatically inherited the observatory through her.

An even closer parallel to Elisabetha Hevelius is Maria Winkelmann. She married Gottfried Kirch, an astronomer who had been trained by Johannes Hevelius. Working with her husband, Maria Kirch made original observations and taught practical astronomy to her own children. Even though she may never have met Elisabetha Hevelius, she must have known about the older woman's expertise and active involvement in the Startown observatory.

Although there are no surviving records of comparable astronomical dynasties in England, the skilled craftsmen who made astronomical instruments similarly handed down their businesses through the generations. Women shared in the family work, and their power is made particularly evident by examining inheritance struggles – women become visible because of their legal rights. A famous example is the Adams family, who ran one of London's most prestigious instrument concerns for almost a hundred years. When the first George Adams died in 1782, his wife Ann (her original surname is lost) took over the shop in Fleet Street until her son George Junior was fully qualified. Ann Adams was still alive when the younger George Adams died twelve years later, but within a fortnight her daughter-in-law – George Junior's wife Hannah Marsham – had staked her own claim to the profits by trying to sidestep his brother Dudley. Dudley and his mother managed to force the wife out, but this proved to be an unwise strategy: the well-organised wife, Hannah Marsham, left a substantial estate of £20,000, whereas Dudley Adams bankrupted the family business, later becoming a shady political agitator and electrical therapist.[11]

In comparison with England, the guild tradition remained strong in Germany long after the mid-seventeenth century, and it seems to have fostered female participation in astronomy. Although women found it hard to gain formal recognition, they could claim an important, if secondary, role in astronomical households because making instruments was a skilled craft. Traditionally, astronomical careers were founded on training with a master inside his private observatory, not by following academic courses in a university. Women were regularly taken on as apprentices, or inherited family businesses from their husbands or fathers. They fulfilled vital functions. Like Jane Dee, they had to run the household and look after all the apprentices and hired assistants. In addition, they participated in the astronomical observations, sometimes staying up all night to record star positions and then working during the day to carry out mathematical calculations and catalogue the data.

Unsurprisingly, participation did not mean equality. In 1721, one eminent member of the Berlin Academy boasted that the high number of female astronomers in Germany put the rest of Europe to shame. Their accomplishments, he proclaimed, 'make one recognise that there is no branch of Science in which Women are not capable'. But he was in a minority. He protested about his prejudiced colleagues who resented Maria Kirch simply because she was a woman: '& they would have liked to relegate her to her distaff & her spindle,' he fumed ineffectively.[12] A generation younger than Elisabetha Hevelius, and related to her via Johannes Hevelius's assistant Gottfried Kirch, Maria Kirch illustrates how women were squeezed out of professional astronomy. Her story shows how their achievements could be masked behind the reputations of their fathers and husbands.

Like Catherina Koopman (later Elisabetha Hevelius), Maria Winkelmann (later Maria Kirch) (1670–1720) was interested in astronomy as a child. But she was realistic: university was not an option because she was a woman. Her first step was to become the assistant of a local farmer, who had stunned the elite astronomical world by reporting a new comet eight days before Johannes Hevelius found it. (Like astronomical women, this so-called 'Peasant' initially published anonymously.) Then Winkelmann adopted Koopman's own strategy – marry an elderly distinguished astronomer whose first wife had died. Her family preferred a more orthodox candidate, a young Lutheran minister, but Winkelmann insisted on marrying Gottfried Kirch, Germany's most famous astronomer.

The contrasting educational experiences of these astronomical partners were typical. As a man, he had benefited from two types of teaching: traditional training in practical astronomy at Startown in Johannes Hevelius's workshop and observatory; and the relatively recent scholarly practice of studying mathematics at university. As a woman, Winkelmann relinquished her unofficial apprenticeship under one man to become another man's wife and assistant. She acquired status, but she also acquired domestic responsibilities, including caring for her own children as well as those of her predecessor.

Elisabetha and Johannes Hevelius, Maria and Gottfried Kirch: these scientific couples worked as partnerships, as teams. There was no hard boundary between home and observatory, and the wife's role merged into the husband's. Sometimes Maria and Gottfried Kirch worked side-by-side, but divided the sky between them; that was why Maria Kirch was the first to observe the northern lights, when her husband was watching the southern half of the sky. At other times, they collaborated when two observers were needed, or else took turns so that between them they could cover the entire night.

And it was on one of those occasions, while her husband slept, that Maria Kirch spotted a comet that her husband had missed. She was the only astronomer in the whole of Germany to see it, and for the next couple of weeks they tracked its course together. So whose discovery was it – hers or his? The official scholarly language was Latin, but Maria Kirch only spoke German. Gottfried Kirch wrote the first reports in Latin, and Gottfried Kirch got the credit. It was only eight years later, in the first volume of the new journal at Leibniz's Berlin Academy, that he opened an article with the words 'my wife . . . beheld an unexpected comet'. Over the decade that they observed together, Maria Kirch's observations were generally concealed behind her husband's façade.[13]

Together in Berlin, Maria and Gottfried Kirch manoeuvred to get more funding for astronomy. She petitioned Leibniz for support, and he used his friendship with Princess Sophie to present her at court. 'I don't believe,' Leibniz enthused, 'that this woman could easily find an equal in the science at which she excels . . . she observes as well as the best [male] observer.'

And to placate Sophie's doubts, he assured her not only that Maria Kirch was technically competent with instruments, but also that she was as well versed in the Bible as in the fashionable Copernican idea that the sun is stationary at the centre of our planetary system.[14]

But even with Leibniz as her patron, Maria Kirch could not protect her position at the Academy after her husband died. Although she had worked by his side on the Academy's calendars for the past ten years, she lost her place as his assistant and was about to be turned out of his official home. She was desperately worried about supporting herself and her four children. Although she applied for an inferior job at the Academy, the Secretary fretted about setting a precedent by employing a woman. 'Already during her husband's lifetime,' he wrote to Leibniz, 'the Society was burdened with ridicule because its calendar was prepared by a woman. If she were now to be kept on in such a capacity, mouths would gape even wider.'[15] Leibniz rallied to her support, but won only a brief reprieve: six months' housing, about a month's wages and a gold medal.

Maria Kirch survived by moving to a private observatory, where she rose to become a master astronomer with two students of her own to train – under the craft guild tradition, women could achieve high positions in Germany. The Hevelius family later invited her to reorganise the Startown observatory, and she stayed there for eighteen months. But academic astronomy excluded women. Even though she did eventually publish in German under her own name, Maria Kirch could only get back inside the Berlin Academy by becoming assistant to her own son Christfrid.

This pattern of discrimination was perpetuated. Like his father, Christfrid Kirch had benefited from the dual astronomical education that was unavailable to his mother: a year at university as well as practical training in observation from his parents. And so Christfrid Kirch inherited his father's position at the Academy, the job that was denied to his mother. Maria Kirch's daughters had received the same domestic apprenticeship as their brother, but were not eligible to go to university; echoing her experiences, they could only find paid work as Christfrid Kirch's assistants. After he died, they were forced to observe from home, where they complained about the tiny windows and their inferior instruments.

Leibniz told Princess Sophie that Maria Kirch was 'a most learned woman who could pass as a rarity . . . one of the best rarities in Berlin.'[16]

What a double-edged compliment! Was she a priceless treasure to put on display, or was she a curiosity, a freak of nature to be marvelled at? In any case, as Leibniz knew, although she might have been rare, she was not unique. Three years earlier, a German pastor called Johann Eberti had published his *Open Cabinet of Learned Women*, an influential set of female biographies. Trawling through all the available literature, Eberti assembled together almost 500 exceptional women from the past, sorting them by his own criteria into saints, heroines and scholars.

Like pinning out butterflies in a case, displaying rare women within a collection was not the same as granting them freedom in the real world. Eberti's *Open Cabinet* provided a valuable resource for men who were keen to parade their egalitarian principles while maintaining their privileged status. One of the women preserved in print by Eberti was Maria Cunitz, who spoke seven languages, married her astronomy teacher and published a book on practical and theoretical astronomy. 'She was so deeply engaged in astronomical speculation,' reported Eberti, 'that she neglected her household. The daylight hours she spent, for the most part, in bed (concerning which all manners of ridiculous events have been reported) because she had tired herself from watching the stars at night.' Later writers trotted out Eberti's account of Cunitz because it confirmed their opinion that being an astronomer was incompatible with a wife's domestic duties.[17] Women like Elisabetha Hevelius, Margaret Flamsteed and Maria Kirch demonstrate their error.

Chapter 8

Caroline Herschel / William Herschel

Is it not then more wise as well as more honourable to move contentedly in the
plain path which Providence has obviously marked out to the sex, and in which
custom has for the most part rationally confirmed them, than to stray awkwardly,
unbecomingly, and unsuccessfully, in a forbidden road? . . . Is the author then
undervaluing her own sex? – No. It is her zeal for their true interests which leads
her to oppose their imaginary rights.

Hannah More, *Strictures on the Modern System of Female Education,* 1799

February 1828: an unusual meeting is taking place at London's Astronomical
Society. The Vice-President, James South, is discussing William Herschel,
the astronomer most famous for discovering the planet Uranus. 'Who
participated in his toils?' asks South in mock-astonishment. 'Who braved
with him the inclemency of the weather? Who shared his privations? A
female – Who was she? His sister.' Although much of his speech is about
William, South is about to award a coveted gold medal to this sister,
Caroline Herschel (1750–1848). (Conveniently, she is far away in Germany,
so South has not had to confront the question of whether she should be
invited into this male enclave to receive her prize.) Caroline Herschel has
discovered eight comets and several nebulae, but South admires her for
the calculations she carried out on William Herschel's observations, for
her 'unconquerable industry' in helping her brother.[1]

February 1835: seven years later, and the Royal Astronomical Society is in
a quandary. Should they make Caroline Herschel an honorary member,

even though she is a woman? After much debate, they vote to admit her, arguing that old-fashioned prejudice should not stand in the way of paying tribute to achievement. They formulate what is perhaps the first statement of equal opportunities in science: 'while the tests of astronomical merit should in no case be applied to the works of a woman less severely than to those of a man, the sex of the former should no longer be an obstacle to her receiving any acknowledgement which might be held due to the latter'.[2] The language may be outdated, but the sentiments are modern.

Caroline Herschel was rewarded not for her own discoveries, but because she recorded, compiled and recalculated her brother's observations. She was 'his amanuensis', reported South, thus consigning her to the scientific margins. South described how, after a night's work together at the telescope, William observing and Caroline recording, it was Caroline who spent the morning copying out the readings neatly. And it was Caroline who performed the calculations, arranged everything systematically, planned the next day's schedule and collated their observations in publishable form.

In some ways, Caroline and William Herschel's astronomical partnership resembles the family craft traditions in Germany a hundred years earlier. Like Elisabetha Hevelius and Maria Kirch, Caroline Herschel was acknowledged to be highly competent, but her achievements were concealed behind the name of the man in the family. Yet there were also significant differences. Most obviously perhaps, the Herschel family was known for its musical rather than its astronomical expertise; William Herschel did not undergo a normal apprenticeship and never studied at university. Just as importantly, by the end of the eighteenth century attitudes towards science as well as towards women were changing. In England, enterprising men had shown that science was useful and could also be profitable – earning enough money to support a family now seemed a viable option. Campaigners like Mary Wollstonecraft were demanding better education for girls, but even women who worked still maintained that the sexes were cut out for separate roles in life.

Caroline Herschel was often described as a humble assistant. Pictures like the one reproduced as Figure 22 consolidated her subservient role. It shows her meekly proffering a cup of tea to sustain William, who scarcely seems to notice her presence as he busily polishes a mirror. The scene radiates domestic harmony. Dressed in a pretty pink gown, she serves while he works. No indication here of her diminutive size and the facial scars left over from a childhood bout of smallpox. One thing jars: the open box on the floor with a miniature doll inside. Might this symbolise Caroline, peering out from the confines of her upbringing, but unable to escape? This romanticised image encapsulates the conventional story of the Herschel couple . . .

Fig. 22
Caroline and William Herschel.
A. Diethe, coloured lithograph.

Originally from Germany, William was a professional musician in Bath when he brought his younger sister over from Germany to train her as an opera singer. As William's hobby of astronomy turned into a passion, Caroline abandoned her musical career and dedicated herself to being his assistant as well as his housekeeper. William's prowess is chronicled by the instruments arranged on the bookcase. Although they are not accurately drawn, they are meant to provide a potted history of astronomy. On the left is an armillary sphere, developed by the Greeks for observing the stars and still used for teaching in the Herschels' time. Next to it, the small brass telescope indicates how William worked when he first became interested in the stars. On the right sits a model of his most famous instrument, a giant telescope housed in a wooden frame.

William succeeded because he was a skilful researcher who constructed instruments of extremely high technical quality. By building telescopes with an extraordinarily high magnifying power, he discovered that some stars were in fact double. In order to see further and further out into space, he needed telescopes that would gather lots of light to detect faint and distant stars. This meant building massive instruments with very large, smooth reflecting mirrors, which William ground and polished himself. Caroline later recalled that he once spent sixteen hours continuously working on a particular mirror, and this picture reflects his reputation as a skilled craftsman who sold telescopes all over the world.

At first, professional astronomers were scathing. A lunatic fit only for Bedlam, muttered conservative critics, who ridiculed this provincial upstart's claims that he could build telescopes twenty times more powerful than existing ones. But William was vindicated in 1781, when he noticed a mysterious object in the sky. Within months, astronomers all over the world were celebrating his discovery of an eighth planet. Like Galileo and Hevelius, he solicited royal patronage by naming it after George III, even though it was soon rechristened Uranus. This was the first modern addition to the classical universe, and the poet John Keats later evoked the keen excitement:

> Then felt I like some watcher of the skies
> When a new planet swims into his ken . . . [3]

After George III invited William up to London to demonstrate his powerful telescopes, the Herschels' life was transformed. Granted a royal pension to provide astronomical entertainment for the King and his family, William could afford to give up the music that had financed his research. Caroline and William moved across the country in order to be near the castle at Windsor, and William built larger and larger telescopes, culminating in one forty feet long with mirrors four feet in diameter.

William's major interest lay not in the planets, but in the stars. Like a natural historian gathering specimens, he wanted to collect stars from all over space and group them into different types. He set out to make an inventory of the universe and map the dimensions of our galaxy. Through his powerful telescopes, what looked like luminous blurs in smaller instruments resolved themselves into myriad star clusters that could be distinguished from genuine cloudy nebulae. Because he had discovered and catalogued distant stars and nebulae, awards for William poured in – honorary degrees, a knighthood, the presidency of the Astronomical Society.

Recognition for Caroline came only after William's death. By then, she was living on her own, forced by William's marriage to move out of the home and workplace they had shared for sixteen years. Following a decision she always regretted, Caroline went back to Germany, where she dedicated herself to helping William's substitute – his son, who became the famous Victorian astronomer, John Herschel. After a life of scientific servitude to her brother, she devotedly compiled star catalogues for her nephew.

By the time that she was fifty, Caroline Herschel had become a token scientific woman, astonishment being reserved for her sex rather than her discoveries. When King George III demanded to see her comet, one of the court attendants – the novelist Fanny Burney – interrupted her game of cards to dash out into the garden and climb up the telescope's steps. 'It is the first lady's comet,' she wrote, 'and I was very desirous to see it.'[4]

Modern biographers have also seized on this 'first', seeking to rescue Caroline Herschel from her secondary role as a scientific helpmate and convert her into a female icon of science. Feminists have rewritten Caroline Herschel's story to underline her independent achievements and the contributions she made towards breaking down prejudice against scientific women. But when does a shift of emphasis become an exaggeration, a distortion? Scientific women have been concealed for so long that it's very tempting to overstate the case and convert them into unsung heroines. Retelling women's stories to make them conform with modern ideals is historically insensitive; moreover, it is not very helpful for understanding how the past has led to the present.

A Victorian feminist interpreted their relationship rather differently. Writing in an anthropology journal, she declared that 'the splendid renown attached to Sir W. Herschel's name was largely due to his sister's superior intelligence, unremitting zeal, and systematic method of arrangement'.[5] Caroline Herschel's 'superior intelligence' may be hard to prove, but this claim that William Herschel's fame depended on her activities was no exaggeration. Far more than a simple helper, she was indispensable for establishing William Herschel's reputation and compiling his work for publication. Through her collaboration with her brother, Caroline Herschel strongly affected the course of astronomy. This aspect of her achievements seems far more significant than her discoveries of a few small comets, several of which were in any case also observed by other astronomers.

When astronomers are observing the night skies, they often dictate readings to a scribe so that they can keep their eyes attuned to the dark. William Herschel ran a family-based operation. His major assistants were his sister Caroline, who converted raw data into publishable, error-free knowledge, and their brother Alexander, a musician in Bath who was probably as technically gifted as William. Beneath them in the hierarchy were the men who manipulated the heavy telescopes at night. And still lower were the fifty or so artisans, identified by numbered shirts, who carried out the unskilled manual labour. All of these invisible assistants have been concealed by William Herschel's reputation.

Caroline Herschel's role was exceptionally important because of the very nature of her brother's research. Unlike previous astronomers, he concentrated on collecting and classifying large numbers of stars, so that Caroline Herschel's cataloguing skills were of central importance. It was after her early success at logging nebulae that William Herschel decided to pursue this new line of enquiry, the one that came to dominate his research and consolidate his fame. Lacking her patience and meticulous attention to detail, he relied on Caroline Herschel's scientific input. Joined by their brother Alexander in the summer months, they worked together as a team – although only William Herschel received the prestige.

Even sympathetic biographers are forced to recognise that Caroline Herschel colluded in her downtrodden state. 'I am nothing, I have done nothing,' she wrote; 'a well-trained puppy-dog would have done as much' – a self-abnegating remark that cannot simply be dismissed. Because Figure 22 demotes her into a tea-bearing servant, feminist campaigners would like to dismiss it as a typical chauvinistic denigration of women's involvement in science. However, while it does serve this function, it also matches Caroline's own description of the care she lavished on her 'brother when polishing, since by way of keeping him alive I was constantly obliged to feed him by putting the victuals by bits in his mouth . . . serving tea and supper without interrupting the work with which he was engaged.'[6]

Commentators coined polite euphemisms to disguise Caroline Herschel's slavish servility. 'All shyness and virgin modesty,' remarked Burney's father when he met her, and Burney herself acutely noticed how Caroline Herschel allowed her brother to answer questions on her behalf. Humility is itself a form of greatness, judged a sanctimonious Victorian relative keen to polish up the family reputation. A 1930s biographer praised 'her beautiful character', even though Caroline Herschel's autobiography reveals an embittered, prickly, solitary woman who shunned friendly advances and perceived the world as a hostile place.[7] Recent writers, keen to promote female participation in science, have glossed over her emotional difficulties to focus on her achievements. Concealing her problems seems almost as insulting as minimising her ability, and also reduces the modern relevance of her experiences. Many women still lack self-confidence and allow themselves to become victims of a scientific system that is geared towards aggressive, arrogant styles of behaviour.

Caroline Herschel was evidently a woman of huge ability, who did make independent discoveries, but rewriting her life prompts important questions about how science's own history should be told. Does it make sense to celebrate her eight comets rather than the many years she devoted to systematic research? Methodical work lies at the core of scientific progress, yet we still celebrate the unusual, the breakthrough, the single spectacular event. Caroline Herschel's importance for her brother's research was far greater than most of their contemporaries realised, yet she allowed her reputation to be channelled through him. How did this happen? Was it just a consequence of eighteenth-century attitudes towards women, or do her experiences have resonances in modern science? Caroline Herschel's experiences illustrate how compromises, concessions and survival strategies can have damaging long-term effects, as a person gradually slides into a situation of no-return.

Caroline Herschel was immured in domesticity throughout her life. Like a failed Cinderella, she escaped the drudgery imposed by her mother and her family, only to become the handmaiden of her elder brother, the adored yet demanding fairy prince who whisked her off to England. Under his care, her duties expanded beyond the conventional cleaning and cooking to include menial tasks like sieving horse manure to make smooth beds for telescope mirrors. Even her independent observations relied on routine, as she systematically searched the skies for comets, an activity with an appropriately domestic label – sweeping.

As a child in Hanover, her father – an army musician – showed her the constellations and wanted her to be educated, but her mother was obdurate. Caroline, she insisted, must stay at home to cook and clean. Illiterate herself, perhaps the mother resented Caroline's attempts to gain skills and opportunities that she herself had been denied; certainly, with six surviving children, she wanted to lighten her own workload by keeping her daughter at home. From her perspective, learning threatened the family harmony by giving her sons high ambitions that would take them away from home, and she squashed heated scientific discussions in case they woke the younger children.

Caroline seethed with resentment and daydreamed unrealistically about life as a governess. 'I could not bear the idea of being turned into an Abigail or housemaid,' she wrote later in a memoir that is choked with self-pity and frustration. She stored up bitter memories: being banned by her mother from dance lessons that had already been paid for, missing family reunions as she toiled in the scullery, harsh punishments for her clumsy serving at dinner parties. She soon learned to take her pleasures vicariously, eavesdropping on her brothers' animated conversations because 'it made *me so happy* to see *them so happy*'.[8] Looking back, she singled out William for his kind behaviour. Twelve years older, he was the only one who comforted her when she returned, cold and forlorn, from an unsuccessful search for her father at a military parade.

Opting for music rather than war, William escaped from the army when Caroline was seven and found refuge in England, where he earned his living by composing, playing and teaching. But he was dissatisfied. Music, he thought, should be for pleasure, not for money. Also, he was lonely. In three years, he complained, 'I have not met a single person whom I could feel worthy of my friendship.' He went home briefly, the popular son who kindly complimented his little sister's confirmation dress, before dashing back to England leaving her plunged in depression, mechanically trudging from one chore to the next. First her father died, then her only friend; her younger brother went to England, but Caroline was left behind, endlessly sewing sheets and knitting ruffles, too proud 'for submitting to take a place as Ladiesmaid' and scarred by her father's advice to forget about marriage 'as I was neither hansom or rich'.[9]

Eight long years went by before she was rescued. Silencing their mother's protests by hiring a family servant, William announced that he was taking Caroline to England to train her as a singer. Torn between joy and guilt, she made enough clothes to last everyone a couple of years, and embarked on the two-week journey to Bath, where she slept in the attic with Alexander, while William occupied the handsomely furnished rooms on the first floor. William had already attained god-like proportions in Caroline's eyes, and she spent the rest of her life revolving around him. As her nephew commented: 'There never lived a human being in whom the idea of Self was so utterly obliterated by a devotion to a venerated object.'[10] Uneducated, knowing no English, Caroline willingly submitted herself to William's domination, binding herself deeper and deeper into a situation of voluntary exploitation.

Her days settled into arduous routine, broken only by periods of relaxation listening to William enthuse about astronomy. Work started over an early breakfast, when William taught her English, arithmetic and book-keeping – the skills she needed to manage the household for him. Then followed her singing practice, squeezed in before her other duties. She started by tackling English cooking and the nightmare of shopping in a foreign language, but soon increased her responsibilities by sacking the servant ('a hot-headed old Welshwoman,' she sniffed).[11] Like many lonely people, she even resented attempts to break her solitude. Friendly visitors were dismissed as idiots, and she bristled with animosity throughout a trip to London, jealously suspecting her hostess of wishing to marry William.

Caroline's musical career flourished under her brother's guidance. She started to be billed in solo parts, but after six years she turned down the offer of a performance in Birmingham that could have been the start of an independent life. 'I never intended to sing anywhere but where my Brother was the Conductor,' she declared.[12] Sisterly loyalty, or self-punitive fear of freedom? Recognising her devoted commitment, William liberated himself by passing on tedious chores to her. As her days became eaten up in training the choir and copying out sheet music, she relinquished all her own ambitions.

William's obsession with astronomy had already started to take over their lives. Over the breakfast table, he imposed 'Little Lessons for Lina' that included enough algebra and geometry for her to be useful in recording observations, and he kept snatching her away from singing to help him test his instruments and build new telescopes. To Caroline's horror, a couple of summers after she arrived in Bath, William converted their home into a workshop. A carpenter took over the elegant drawing-room, a large lathe was installed in a spare bedroom and soon the garden became a construction site. William was still giving concerts and lessons, even though his pupils were disconcerted by the globes and telescopes piled up on the piano.

Caroline was not just a passive helper. Down in the basement, while a furnace melted the metals to cast enormous mirrors, Caroline spent her days making the mould 'prepared from horse dung of which an immense quantity was to be pounded in a morter and sifted through a seaf'. When the mould cracked during casting, she started work on a new one. With no hint of disgust, she merely remarked that 'it was an endless piece of work and served me for many hours' exercise'.[13]

Living and sleeping inside a house dedicated to producing astronomical observations, she 'became in time as useful a member of the workshop as a boy might be to his master in the first year of his apprenticeship'

– Caroline's own words. Given her propensity to self-denigration, it seems clear that she worked far harder than any paid apprentice. As well as menial tasks, she was in charge of William's notebook, crossing through the roughly written pages when she had made a neat copy. When Bath emptied during the summers, there was more time available for making instruments, and Caroline's work intensified. Her duties, she remarked, 'left me no time to take care of myself or to stand upon nicetys', suggesting that she retreated from the world as William made more and more demands on her time.[14]

Systematic searching with a powerful telescope: that was William's strategy, and on 13 March 1781 it paid off. Afterwards, he claimed that he would inevitably have discovered the planet Uranus sooner or later; like reading a book, he wrote, he simply turned the page that had the seventh planet recorded on it. But at the time he failed to recognise what he was looking at, and Caroline dutifully copied out his scrawled note of 'a curious either Nebulous Star or perhaps a Comet'. In London, the Fellows of the Royal Society were stunned to learn about this provincial astronomer and his massive telescope, and Bath tourists started to include the Herschel household in their itinerary. When it became clear that this strange new star was a planet, William's astronomical allies started negotiating on his behalf, and he was summoned up to London to meet George III. Like Caroline, he disliked being wrenched out of their private world. 'Company is not always pleasing,' he complained to her, 'and I would much rather be polishing a speculum [a metallic mirror used in telescopes].'[15] But he must have made a good impression, since he was awarded a royal pension so that he could give up teaching music and devote himself to astronomy.

Astronomy was William's major priority, so he chose a house that was near Windsor and had outbuildings that could easily be converted. Caroline was appalled when she saw it. The roofs let in rain, the weeds were four feet high, and even getting there was difficult. Visitors preferred to arrive on Sundays, when the highwaymen were deterred by the large numbers of carriages flocking back to London after a weekend in the country. She soon discovered that their pension would not cover the high local prices, and money became a constant source of anxiety. As she virtuously economised, she came to perceive a world plagued by grasping, uncooperative people.

Self-sacrifice slid into deliberate self-abasement as she prided herself on going without all but the most basic necessities. One of her few personal possessions was a worry-bead.

Alexander went back to Bath to continue his musical career, but Caroline decided that there was no turning back. She and William (with Alexander in the summers) spent every spare moment making mirrors and telescopes, many of which they sold, each with an instruction booklet carefully copied out by Caroline. Polishing the large mirrors was a time-consuming job, and for hours on end she docilely read novels to William, or popped food into his mouth so that he could work without interruption. With no time to change their clothes, 'many a lace ruffle was torn or bespattered by molten pitch, &c.' – presumably more housework for Caroline and the servants she had such difficulty in keeping.[16] She stayed up with William at night, bringing coffee to keep them awake and busying herself in making measurements, copying out catalogues, tables and papers – sometimes in Latin – and helping to adjust the telescope's position. They worked together in the dark, even when the ground was deep in snow. Plummeting temperatures sometimes froze the ink and once even cracked the metal mirror. 'I know how feverish and wretched one feels after 2 or 3 &c nights waking,' she later commiserated with her astronomer nephew.[17]

The full significance of her sacrifice hit her hard, and she consoled herself with work, at her brother's beck and call 'either to run to the Clocks, writing down a memorandum, fetching and carrying instruments, or measuring the ground with poles, &c., &c., of which something of the kind every moment would occur . . . In my leisure hours I ground 7 feet and plain mirrors from ruff to fining down and was indulged with polishing and the last finishing of a very beautiful mirror'.[18] Perversely, she seems to have salvaged her pride by revelling in her own abnegation, even worrying about disrupting the observation schedule when, stumbling around in the dark on snow-covered ground, she fell over and tore open her leg with a meat-hook.

Instead of practising music, William and Caroline now worked hard at training themselves to look at the stars. Astronomers must be dedicated, proclaimed William – just like learning how to play Handel's fugues on the organ. Music stands even provided the model for his telescope supports. He gave Caroline a small telescope of her own, a little sweeper that, by covering large sections of the sky at once, was ideal for comet-searching. She steeled herself to undertake solitary observations. Although she boasted about her fortitude when helping William, at first she was overwhelmed by 'spending the starlight nights on a grass-plot covered by dew or hoar frost without a human being near enough to be within call'. But she persevered, becoming an expert who could recognise stars and nebulae by sight and so immediately detect any new comet that appeared. William rewarded her with a better instrument, although she was often kept so busy helping him that she had little time for her own research. But however much she complained about his interruptions, Caroline liked the security of knowing that her brother was close at hand. She converted herself into 'a being that *can* and *will* execute his commands with the quickness of lightening.'[19]

A French traveller, entranced by the dream-like harmony of the strange household, stayed up with them all night and described their shared ritual of observation. Caroline sat quietly by the open window, studying an astronomical atlas, yet keeping her eye on a pendulum clock and a special dial linked by strings to the large telescope outside. William was perched on a platform at the top end of the telescope, which was slowly raised and lowered by a servant sitting underneath it. Freezing in the night air, Caroline noted down William's shouted readings so that she could mark the stars on a chart, and yelled back instructions. 'Brother, search towards the star *Gamma*' she might say before resuming her work, and a subordinate would turn the telescope in a different direction.[20] And when the morning came, she might grab a short sleep before continuing with her mathematical work, translating the numbers on her dial into degrees and minutes so that the stars' locations could be mapped precisely.

Life was not always so tranquil. Orders for telescopes kept pouring in, and they worked desperately hard to fulfil them. After William became severely ill from long winter nights spent observing in the damp Thames air, they moved again, this time to Slough, where William immediately antagonised the neighbours by cutting down all the trees so that he could see the stars. Caroline suspiciously regarded all the local workers as thieves. Once again, the garden became a building site while a giant telescope, five feet across inside, was erected. It incorporated a special hut for Caroline, so that she could communicate with William through a speaking-tube. As it lay on the ground before being assembled, George III guided the Archbishop of Canterbury through the tube, exclaiming, 'Come, my Lord Bishop, I will show you the way to Heaven!'[21]

Three months later, William had to deliver a telescope to Germany, and Caroline was left in charge. How should she fill the unaccustomed hours? At first, she diverted herself by polishing all the telescope brass. Then there was needlework, shopping and arguments with workmen (she knew they called her 'stingy'). The next few weeks were spent in sweeping the skies, systematically recording her observations and carrying out the calculations needed to catalogue them properly. Boring, essential work, the sort on which she thrived, despite her moans.

But one night the bland routine was interrupted. Disguising her excitement, she fired off a letter to the Secretary of the Royal Society. 'I venture to trouble you,'she wrote timidly, 'with the following imperfect account of a Comet . . .' As the news spread, astronomers inspected her comet and wrote to her with warm congratulations for this unaided discovery. 'I am more pleased than you can well conceive that *you* have made it,' enthused one; 'you deserved such a reward . . . for your assiduity in the business of astronomy.'[22]

When William came back, she resumed her satellite existence, revolving around her brother to promote his interests. His fame rested on Caroline's accurate recordings of thousands of nebulae. As befits a scientific hero, he gained the reputation of being a selfless and charming man, but Caroline knew how his temper could flare up. A good publicity agent, she learned how to avoid angering him and managed his public image. While he was preoccupied with building yet another giant telescope, she controlled the financial arrangements. She also prepared his articles for publication, compiled lists of stars and carried out all the necessary calculations to complete unfinished catalogues. At night, they observed togther, Caroline by her open window, her brother at the viewing end of the telescope. And when visitors arrived, it was often Caroline who conducted the guided tour of the telescopes.

After fifteen years of servitude, Caroline did, at last, attain some sort of independence. For one thing, she was awarded an official salary from the King of fifty pounds a year to be William's astronomical assistant. The arrival of the first quarterly payment was an important moment for her: 'the first money in all my lifetime I ever thought myself at liberty to spend to my own liking'.[23] In England, unlike France, even men were rarely paid for scientific research, and Caroline Herschel was probably the first woman ever to be in salaried scientific employment.

Independence brought its drawbacks. In her autobiography she left only a terse note that she was to lose her job as William's housekeeper. She tore up ten years of bitter comments, but although her lines no longer exist, it's easy to read between them: after William got married, Caroline was intensely jealous of her new sister-in-law. She demanded an independent salary during the marriage negotiations, which proved very tricky. Caroline rejected William's financial support, his prospective bride worried about competing with astronomy, and he was reluctant to give up such a well-trained, indefatigable assistant.

Eventually a compromise was reached: his wife moved in with William, while Caroline lived in her own rooms above the workshops, later moving out to separate lodgings. Gossip circulated about the new Mrs Herschel's wealth. When William took both women to a concert at Windsor, Fanny Burney snidely reported that 'His wife seems good-natured; she was rich too! and astronomers are as able as other men to discern that gold can glitter as well as stars.'[24] Caroline eradicated her own opinion from the record, but this unwanted liberation did leave her free for her own astronomical research, and she discovered eight comets, as well as several nebulae and star clusters.

William's previous absence had enabled her to discover her first comet, and she immediately started sweeping the skies for more, especially when he was away travelling with his wife. A few months after the wedding, Caroline found her second comet, and immediately notified the Astronomer Royal, who sent her a long breezy letter in return. As she continued to find comets, her reputation spread. 'Thus you see,' marvelled the Astronomer Royal, 'wherever she sweeps in fine weather nothing can escape her.' Although William often acted as intermediary, she also engaged in her own extensive correspondence with distinguished astronomers. By the seventh comet, she had even seemed to gain some confidence, informing the President of the Royal Society that 'As the appearance of one of these objects is almost become a novelty, I flatter myself that this intelligence will not be uninteresting to astronomers.'[25] Although she never entered the Royal Society's buildings, three of her letters were published in the *Philosophical Transactions* – the first to appear under a woman's name.

Caroline's discoveries seemed more startling then than now not only because they had been made by a woman, but also because comets were very significant astronomical objects at the end of the eighteenth century. Many people still believed that comets were sent by God, direct messages of his anger at a sinful world. Mary Shelley boasted about her auspicious birth immediately after one of Caroline Herschel's comets had blazed in the sky.[26] Astronomers denounced such beliefs as irrational superstition: they were keen to demonstrate their own scientific superiority by showing how comets followed precise mathematical laws. Émilie du Châtelet's translation of Newton's *Principia* had appeared immediately after French mathematicians had successfully predicted the return of Halley's comet in 1759, but more evidence about the behaviour of comets was still needed to confirm the theoretical work. All over Europe, astronomers observed Caroline Herschel's comets in order to calculate their orbits. The director of Paris's Royal Observatory, who had met her on a visit to Slough and was renowned for his own comet discoveries, wrote promising to keep tracking her comet and send over his results. 'She will soon be the great Comet finder,' enthused one of William Herschel's friends, who chauvinistically hoped that she would beat the Parisian observers.

Joseph Lalande, one of France's most famous astronomers, was exceptionally supportive. 'At the moment,' he wrote to her, 'comets are what interest astronomers most. We are expecting several from you.' Unlike his English counterparts, Lalande actively promoted women astronomers. He was still mourning the death of Caroline Herschel's French equivalent, Hortense Lepaute, who had performed many of the laborious calculations needed for the 1759 prediction. Lalande christened a new baby Caroline (sister to Isaac, Newton's namesake), reporting in a national gazette that he couldn't have chosen a more illustrious name. In his book on women's astronomy (which was translated into English), he reminded his male colleagues that women's 'abilities are not inferior, even to those of our sex who have attained the greatest celebrity in the sciences.'[27]

Despite her international renown, Caroline was still willing to take on tedious tasks for William. The major British star catalogue had been published seventy years earlier under the name of Johannes Hevelius's rival John Flamsteed, but there were many mistakes – some stars were missing, while others were recorded with the wrong brightness. Caroline was the obvious candidate for the demanding, boring task of correcting the errors. She had already compiled much missing data, but now William suggested that she draw up an Index to make it easy to refer from the catalogue to Flamsteed's original observations. Twenty months of hard, tedious work. Did she know that Flamsteed's wife Margaret Cooke had, like her, been involved in this catalogue? And that Elisabetha Hevelius had finalised and published her husband's atlas?[28]

Caroline might have used William's marriage to prise her freedom, but, by now conditioned into servitude, she found herself unable to escape. Although she was making independent discoveries, in other ways she was more enslaved than ever. On the surface, William's two women overcame their initial hostility to become very friendly, yet Caroline's diaries imply a different story. For twenty years she moved between the observatory at Slough, a succession of lodgings and Alexander's house at Bath, a trajectory that seems as though it were designed to ensure that she encounter her sister-in-law as little as possible.

While William and his wife travelled round the country visiting friends, Caroline was left behind to earn her salary. Rather than describing the purposeful life of an independent woman, her diary entries suggest the haphazard motion of a subordinate who travelled round England to look after empty houses and cope during astronomical crises. One year, she went to Bath in July and came back to William's house in November; after dinner on the first evening, she started work copying out his latest astronomical article. About a week later, the rest of the family went to Bath, leaving her alone until they came back; the next day, she decamped to a nephew's house. The opening entry for 1813 gives the flavour of her life: 'The three last months of the preceding year I spent mostly in solitude at home, except when I was wanted to assist my brother at night or in the library.'[29]

By 1817, William was so ill, old and depressed that Caroline 'felt my only friend and adviser was lost to me'. She slogged on, sorting out papers, helping her brother when he was well enough to work, but more often watching in dismay as his hands trembled over the backgammon board. While the family was away in Bath, she moved in to make an inventory, but when William got back he showed no interest in her efforts. As he declined, she made herself sick with worry, returning each day 'to my solitary and cheerless home with increased anxiety for each following day'.[30] What should she do, she fretted, where should she go? William lingered on until 1822, and by then she had made up her mind – she would return to Germany.

She left in a great rush, not even waiting until all the funeral arrangements were completed. Did she panic, stunned by William's death? But it was hardly unexpected. Even though the courteous correspondence apparently reveals devoted relatives, her decision to depart so abruptly does suggest unbearable family tensions. Certainly she was unhappy once she got back to Hanover. Right from the beginning, she rarely went out, 'and what little I have seen of Hannover . . . I dislike!' Still, she told her nephew John, even this dreary existence was better than staying in England 'where I should have had to bevail [sic] my inability of making myself useful'.[31] Her letters to her sister-in-law and her nephew are masterpieces of emotional blackmail: a pity I haven't heard from you yet, I feel so ill and lonely here, my relatives are such nasty people, if only I was in the fresh English air . . . for twenty five years Caroline sent a constant stream of letters back to England, packed with complaints and regrets. Guilt-tripping is a modern expression, but the concept existed long ago.

She transferred her adoration to William's son John, making him the new centre of her life, and justifying her own existence solely through her usefulness to him. She sifted through the English newspapers – two months old by the time she got them – for any mention of his name, sent him books and papers, repeatedly burdened him with the information that he was her only purpose for living. Suffering from failing eyesight (see Figure 23), and hemmed in by tall buildings, she could not use the telescope that she had installed in her rooms as a monument to the past. Instead, she embarked on yet another catalogue, this time of the nebulae and star clusters that William had observed, intending her work not for publication but for John's personal use to advance his own astronomical career. Day after day, she worked her way through 2,500 sets of calculations. A labour of love, to be sure, and extremely useful, but also the means for an obsessive, unhappy old woman to obliterate her surroundings and survive vicariously

Fig. 23
Caroline Herschel.
Etching by G. Busse,
Hanover, 1847.

through the men in her family. She seemed determined to be miserable, reflecting only on the past and even regretting John's achievements in case they detracted from William's fame. In the autobiography she started when she was eighty, the very first sentence is about William.

Almost a hundred when she died, Caroline became celebrated for her longevity as well as her achievements. She received official awards from London's Royal Astronomical Society and the Royal Irish Academy in Dublin. Presumably she felt some pride in this belated recognition, although when she received her medal from London she ungraciously claimed to have felt 'shocked rather than gratified . . . for I know too well how dangerous it is for women to draw too much notice upon themselves'.[32] When she was ninety-six, the King of Prussia sent her a gold medal, and for her next birthday the Crown Prince and his family arrived to present her with a large velvet armchair. Since she entertained her guests with a song by William, it seems likely that her beloved brother formed a major topic of conversation.

Soon afterwards, she insisted on leaving the comfort of her sofa to pose upright for the drawing shown in Figure 23. Unsurprisingly, this self-effacing woman left very few portraits, but this one pays tribute to her continued interest in astronomy. As she points to a map of the solar system, her finger rests on the gap between Mars and Jupiter, a zone of great interest to contemporary German astronomers. Like her, they focused on picking out unexpected objects absent from the star catalogues, and were trying to distinguish between comets and small planets. With her failing eyesight, did she realise that Uranus lies way beyond the edge of the paper? For once, the picture is about her, not William.

Caroline Herschel always claimed that she undertook her chores solely for her men's benefit, but occasionally she let a glimmer of self-satisfaction escape. Perhaps, she once remarked, she deserved recognition for her 'perseverance and exertions beyond female strength'. Another time, she allowed herself a rare expression of pique at male oppression in a letter of thanks to the Astronomer Royal for arranging the Royal Society's publication of her Flamsteed Index. 'Your having thought it worthy of the press has flattered my vanity not a little,' she started politely. But this time her gratitude came with a sting in the tail: 'You see, sir, I do owe myself to be vain, because I would not wish to be singular; and was there ever a woman without vanity? or a man either? only with this difference, that among gentlemen the commodity is generally styled ambition.'[33]

There was Caroline's dilemma – she wanted to blend in, to be like other women, yet at the same time she contemptuously shunned their company and involved herself in masculine activities. In some ways, she had escaped the confines of her German upbringing to become an independent, successful astronomer. She earned a scientific salary, was friendly with the most eminent astronomers in England and France, and had published her comet letters and catalogue with the Royal Society. Nevertheless, she was trapped in her conventional belief, ingrained since childhood, that women should be subservient to men. 'All I am, all I know, I owe to my brother,' she stressed repeatedly; 'I am only the tool which he shaped to his use.'[34]

Male astronomers, she wrote, were the huntsmen of science, while she was merely a pointer, eagerly awaiting friendly strokes and pats from her masters. This sporting simile immediately summons up the British gentrified life that surrounded her in the Windsor countryside, and she retained this canine imagery long after she had returned to Germany. She coined various versions of her famous comment that 'I did nothing for my Brother but what a well trained puppy Dog would have done: that is to say – I did what he commanded me.'[35]

This remark makes modern women squirm, yet two centuries ago it carried somewhat different connotations from today. Because animals played crucial economic roles in transport and agriculture, dogs were encountered far more frequently in daily life and, correspondingly, canine imagery appeared more often – to fortify William Herschel against critics, one of his colleagues advised him to 'mind not a few jealous barking puppies'. Dogs and people had not yet, however, accommodated themselves to living together in cities. Hanover, complained Caroline, was plagued by 'the intolerable Newsance of the barking of dogs', relieved only when owners were compelled to muzzle them 'like Bears'.[36]

As Anne Conway's portrait (see Figure 15) illustrates, dogs were important possessions. Men regarded themselves as the pinnacle of God's creation, set on Earth to rule over and tame lesser beings. Like animals – men claimed – women were governed by their passions and needed to be controlled. In pictures, they were shown together, twin models of fidelity and obedience to their master. Mary Wollstonecraft railed against the subservience exhibited by women like Caroline Herschel. 'Considering the length of time that women have been dependent,' she wrote, 'is it surprising that some of them hunger in chains, and fawn like the spaniel?'[37]

Painfully aware of her awkwardness, Caroline was concerned about alienating friendly colleagues. 'I have too little knowledge of the rules of society,' she worried, 'to trust much to my acquitting myself so as to give hope of having made any favourable impressions.' Like many outsiders, she seems almost to have cultivated her strangeness as a defence against a world she perceived as hostile. 'I don't tell *Fibs* though they may not always like to hear what I say,' she wrote defiantly.[38]

Men could cultivate eccentricity as an attribute of genius, and were free to indulge their scientific passion at the expense of normality. In contrast, unusual women were pariahs. 'There is hardly a creature in the world more despicable or more liable to universal ridicule than that of a learned woman,' wrote Lady Mary Wortley Montagu, who was renowned for introducing smallpox inoculation. Clever women were not just breaking the bounds of social convention. They were, it was believed, defying their inherent make-up, and so ran the risk of being regarded as freaks of nature. Montagu was speaking from painful experience when she recommended that her granddaughter should 'conceal whatever Learning she attains, with as much solicitude as she would hide crookedness or lameness'. Lalande once addressed Caroline Herschel as 'Learned miss' in English at the beginning of a letter in French; he did not realise that this could be an insult.[39]

Caroline Herschel avoided the slur of brilliance by moving beyond reticence into taciturnity and converting modesty into self-abnegation. But she suffered. She rejected the friendship of women, most of whom she denigrated as stupid, and so was an outcast amongst her own sex. On the other hand, there was no place for her in the world of men. A fond niece in Germany seemed to be relieved when Caroline Herschel died because, she wrote, 'now the unquiet heart was at rest.'[40]

Chapter 9

Marie Paulze Lavoisier/Antoine Lavoisier

No one has more intelligence, aptitude, and talent for all kinds of work . . . She
[Madame Lavoisier] has a masculine soul in the body of a woman.
Pierre Du Pont de Nemours, letter to Philippe Harmand, 1801

When Judy Chicago laid the table for her imaginary dinner party, she brought together guests from many cultures. But no essential quality of womanhood could have united these visitors. Even scientific women from the same period had very different experiences. Consider Marie Anne Pierrette Paulze Lavoisier and Caroline Herschel, who lived at almost the same time but had only one major characteristic in common. From their childhoods, the lives of both women revolved around one single man. Herschel dedicated herself to her brother, and Paulze Lavoisier to her husband, the famous French chemist Antoine Laurent Lavoisier. There the similarities between these two women end.

To start with, they were born into very different family circumstances and displayed different personalities. Herschel's life was characterised by poverty and self-sacrifice, and she focused her attention inwards, towards the psychological and physical demands of her brother. Scientific gossip or local news did occasionally intrude into her family memoirs, but generally she insulated herself from world events. In comparison, Paulze Lavoisier was rich and outward-looking, deeply committed to social reform, and her fate was governed by the French Revolution. In addition, there were strong contrasts between French and English science. England's prominence relied on private enterprise

167

and individual initiative – the Herschels' small royal salaries were unusual. But in France, both before and after the Revolution, the state supported scientific research financially. As a consequence, science developed differently in the two countries.

Caroline Herschel subsumed her existence beneath her brother's, but Antoine Lavoisier's wife did make a partial declaration of independence. Paulze was her maiden name, and even after she was married, she signed her published chemical drawings and business correspondence Paulze Lavoisier. In recognition of this semi-bid for a separate identity, that will be her name for this chapter.

Paulze Lavoisier (1758–1836) and her husband moved in the same radical circles in Paris as the artist Jacques-Louis David, who painted several of the huge canvases now hanging in the Louvre. Although history paintings of classical scenes later became unfashionable, during the Enlightenment they were regarded as the pinnacle of artistic achievement. By depicting well-known events or myths from the past, they provided object lessons in ethics and taught viewers how to make moral decisions. David proclaimed his revolutionary allegiance by painting huge canvases of Roman scenes that were imbued with modern political significance.

David was also a brilliant portrait artist, and he yielded to the persuasion of Paulze Lavoisier, who was one of his students, to paint her with her husband (Figure 24). This stunning work dominates the gallery where it is on display in New York's Metropolitan Museum of Art. The picture's sheer size – around nine feet by seven – is immediately overwhelming, to say nothing of its shimmering light and colour. The vibrating reflections from the glass vessels, the luminous white cambric of her dress and the polished metal of his shoe buckles all set off the rich glow of the red velvet tablecloth. Even Lavoisier's dark suit has a lustrous texture, accentuated by his white cuffs and jabot, which sparkle like the experimental flasks.[1]

Despite this picture's radical credentials, it became a victim of revolutionary activity. As a wealthy landowner and tax collector, Lavoisier was inevitably a target of suspicion, and his financial activities ultimately led him to the guillotine during the Reign of Terror. He was also in charge of gunpowder production, and in 1789, three weeks after the storming of

Fig. 24
Marie Paulze and her husband Antoine Lavoisier.
Jacques-Louis David, 1788.

the Bastille, a riot erupted when Parisians accused him of shipping powder out of the city to prevent the people from using it. David's picture, completed the previous year and overflowing with optimism for future reform, was diplomatically removed from display and replaced by a less contentious couple whom he also painted, the mythological Greek lovers Helen and Paris.

David intended this double portrait, like his history paintings, to tell
several stories . . .

Most obviously, this picture is about success. When he commissioned
this intimate scene of a fashionable Parisian couple, Lavoisier had paid
David almost five times a professor's annual salary, and the artist produced
an appropriately sumptuous and exquisitely painted canvas. Wealthy by
birth as well as through marriage, Lavoisier funded his comfortable life
from his financial activities as a state tax collector. But his major interest
was chemistry, and he is often said to have initiated a chemical revolution that overturned science as dramatically as the political Revolution
in which he died. Painted in 1788, the year before the Revolution broke
out, David's picture is a dazzling display of bourgeois accomplishment.

Lavoisier's elegant extended leg divides the scene diagonally, so that
Paulze Lavoisier's dominating central presence is balanced by her
husband's chemical equipment and papers, which are all on the right-
hand side. The apparatus carefully arranged on the table refers to
Lavoisier's major achievement, the identification of oxygen. As well as
being scientifically important, this discovery carried great national significance, because it symbolised how Lavoisier had overthrown the older
system of his English rival, Joseph Priestley. At his feet, the large glass
balloon nestling in its plaited horsehair collar was used for collecting
and weighing air. Starting at the bottom right, the instruments have been
ordered chronologically to represent the major stages in Lavoisier's career.

Goose quill in his hand, Lavoisier is correcting the proofs of a book,
probably his *Elements of Chemistry*, which came out the following year. This
was his most important book, in which he persuasively laid out his revolutionary agenda to found a new chemistry. Taking his cue from algebra,
Lavoisier introduced a systematic chemical language – essentially the
names and symbols still in use today. Applying his financial acumen, he
insisted on a balance-sheet approach, expressing reactions as equations
and meticulously checking that weights and volumes always added up.
Lavoisier's new chemistry depended on precise measurements, for which
he designed accurate instruments: hence their prominence in this portrait.

David and the Lavoisiers both moved in the same progressive groups,
and here the politically sensitive artist is advertising the radical chemist's

forward-looking ideas. Even the portrait's colours seem revolutionary. Lavoisier wears Third Estate black, and the canvas is dominated by red and white, laced with the suitably feminised blue of Paulze Lavoisier's sash. And although it shows an elite couple, the workers have not been forgotten: by painting the experimental instruments with the loving attention of a Dutch still-life master, David celebrates not only Lavoisier's ingenuity in conceiving them, but also the skill of the artisans whom he employed to execute them so beautifully.

David is pledging his conviction that future improvement lies in scientific research, a controversial commitment that still needed to be openly declared. Lavoisier boasted about revolutionising chemistry, and he was also an active social reformer. He instigated many administrative changes that made local government more democratic, and he used his scientific expertise to improve people's lives – by renovating Paris's decrepit drainage system, overhauling the dilapidated prisons and hospitals, rationalising the old-fashioned system of weights and measures into the metric system. He also wanted to revolutionise France's agriculture and remedy the chronic food shortages. From his laboratory research, Lavoisier argued that physical labourers required more to eat than sedentary workers, and at his model farm in the country, he analysed his planting experiments with an economist's precision, keeping accounts of the human energy expended as well as the crop yields and selling prices.

David and his colleagues were trying to revitalise the French arts. They turned to antiquity for inspiration, yet at the same time contrasted the old and the new; three classical pilasters dominate the bare back wall of this innovatory study-cum-laboratory. For this modern double portrait, David chose a very similar pose to his treatment of Helen and Paris, another devoted pair. In both pictures, the man looks adoringly up towards his female partner as though they were destined for one another. But whereas Helen submissively returns Paris's captivating gaze, Paulze Lavoisier stares confidently out at the spectator. As foundations for a better society, David was prescribing not only science but also new concepts of love and liberty.

Or was he? David has bequeathed an ambiguous version of Paulze Lavoisier. Lavoisier seems to be beseeching her for guidance, and yet, draped decoratively on his shoulder, she could well be a devoted yet ordinary wife totally ignorant of chemistry. Her deceptively natural coiffure *à l'anglaise*, a powdered blonde wig concealing her own dark hair, suggests long hours dedicated to her appearance. In her simple white dress, the traditional colour symbolising female purity, Paulze Lavoisier might personify female nature. It is hard to tell whether Lavoisier is awestruck by his wife's aura or by the subject he studies. On the other hand, she appears distracted, even irritated. Has she been interrupted in the course of advising her husband about a detail in his proofs? Is she the driving energy behind his fame, or is she merely a figurehead, a muse who provides a source of inspiration with no real knowledge or skills of her own? One of Paulze Lavoisier's friends wrote a verse for her that neatly summarises some of the painting's ambivalence:

> For Lavoisier, subject to your laws
> You fill the double role
> Of muse and secretary.[2]

Still further stories lie concealed within this picture. Separated from the zone of scientific activity by Lavoisier's outstretched leg, Paulze Lavoisier's domain covers the left-hand half of the portrait. On the armchair in the back corner is a large folder, David's tribute to her activities as his pupil. Inside this portfolio could be hidden some of her own pictures – the experimental sketches that document her participation in the research carried out in their own private laboratory at home (Figure 25), and the diagrams she drew for Lavoisier's *Elements of Chemistry* (Figure 26). Paulze Lavoisier left behind firm evidence of the active part she played in building her husband's career.

Marie Paulze was only thirteen when she rejected her first suitor – a man of fifty – and instead agreed to marry Lavoisier. This ambitious young lawyer, twenty-eight years old, was one of her father's business colleagues. Fresh out of a convent, and with only a scanty education, she found herself in charge of a Parisian household, and the wife of a wealthy financier who was obsessed by chemistry and geology. Lavoisier followed a rigorous schedule: science from six to nine in the mornings, a full day of business meetings, then back into the laboratory for three hours after dinner. If she wanted to spend time with him, she had to work. Despite her youth, Paulze Lavoisier immediately embarked on intensive courses of self-instruction so that she could share her husband's activities. For more than twenty years they worked and travelled together.

An English girl in her situation would have found herself automatically excluded from male intellectual circles, and might have idled away her time at parties or devoted herself to her children. But because she lived in Paris, Paulze Lavoisier could enter mixed conversational groups, which were often presided over by women. Although – like Émilie du Châtelet – she could not attend meetings at the Academy of Sciences, she soon found out what had happened there because Lavoisier brought his friends back to tea with her afterwards. Women, according to French Enlightenment ideology, were civilising agents whose intrinsic docility was valuable because it moderated male aggression. Their function was to eliminate brutality and oppression, and so help create a polite, intellectual nation. The eminent naturalist Georges Buffon argued: 'It is only among the nations civilised to the point of politeness that women have obtained that equality of condition, which however is so natural and so necessary to the gentleness of society.'[3]

Women and men were different yet complementary – a condescending view by modern standards, but at least it gave women a central role. By placing Paulze Lavoisier in the centre foreground, close to her husband, David has painted an idealised picture of civilised collaboration. The pile of papers on the table is a product of their joint labour, the book to which they had both made their own individual contributions. Like Lavoisier's own molecular model of chemical affinities, the couple have reached an equilibrium by balancing their opposing characteristics. They form a stable pair, the foundational unit recommended by Jean-Jacques Rousseau for a good society.[4]

At this time, before science had become professionalised, there were no established codes of behaviour for either men or women who were involved in experimental research. Many scientific couples worked together, enjoying marriages that seem to have gained their strength from intellectual companionship rather than from domesticity and parenthood. Like Paulze Lavoisier, the wives of eminent men at the Academy of Sciences often translated foreign scientific books; others wrote novels, plays, political tracts. Several

French women of this period were responsible for introducing foreign texts into France. Marie d'Arconville commented on and translated a major English chemistry text, and Françoise Biot's translation of a German physics book went through four editions. Paulze Lavoisier's friend Claudine Picardet, who also ran an important scientific *salon*, could cope with German, Swedish, Italian and English.[5]

Typically in these couples, the husband was far older, they had relatively small families and both partners engaged in romantic affairs. Paulze Lavoisier and her husband fitted this pattern: they had no children, and Paulze Lavoisier took the opportunity of his frequent absences to become deeply involved with one of his close friends, Pierre Du Pont de Nemours. This was no passing intimacy. They were almost certainly lovers from 1781, and when she rejected his proposal of marriage after Lavoisier's execution, he told his son that 'she had all my heart and she has broken it'. He remained faithful, reminding her much later of 'the inviolable and tender affection that I had vowed to you for thirty-four years'.[6]

French men's lives depended on patronage. By marrying the daughter of a senior colleague, Lavoisier had promoted his legal career and improved his financial situation. Similarly, when the wealthy banker and politician Jacques Necker was entertaining a young relative from Geneva, Albertine de Saussure, Lavoisier consolidated his own position by inviting her round to his laboratory. The daughter of a famous geologist, she had started recording her scientific observations when she was ten years old, and her face was scarred by some early chemical research that had gone wrong. She sent her father excited letters detailing the chemical experiments that Lavoisier had performed especially for her. Later, married to a botanist, de Saussure ran her own *salon*, the scientific counterpart to the literary circle clustered round her cousin, Germaine de Staël.[7] As Lavoisier gained prestige, he started to adopt his own younger protégés – when a junior chemistry teacher wanted to repay Lavoisier for his help, he gave private lessons to Paulze Lavoisier.

The key place for men to meet useful contacts and solicit support was in a woman's *salon*. Like other powerful French wives, Paulze Lavoisier ran a weekly *salon* where men and women met together to discuss the latest gossip, plays and scientific experiments. At her father's house, Paulze Lavoisier had learned how to act as hostess after her mother had died, and so she could rapidly take over this vital role. Within a few years she was leading one of Paris's most important scientific conversation circles, entertaining Benjamin Franklin, Joseph Priestley, James Watt and many other distinguished visitors. Lavoisier's scientific success depended on being able to gain the backing of influential people by inviting them to these successful *salons*, which displayed the intellectual wealth he could gather around him.

Paulze Lavoisier was no mere table decoration – ensuring that these discussion evenings went smoothly was essential for maintaining her husband's position. An American visitor who arrived to negotiate a price with Lavoisier for his Virginia tobacco was surprised that a woman should be more interested in intellectual matters than in the latest fashions. 'She is tolerably handsome,' he remarked, 'but from her Manner it would seem that she thinks her forte is the Understanding rather than the Person.' Nevertheless, he went to the ballet with her, invited himself to tea when Lavoisier was out, and went along with Franklin to one of her evening *salons*.[8] Franklin recognised how powerfully these scientific wives were cementing together a new intellectual community. 'In your company,' he remarked to one *salon* hostess, 'we are not only pleased with you but better pleased with one another and with ourselves.' The *salons* were effectively power bases where wives could arrange patronage for their husbands. They also acted as mother figures for younger men, and Paulze Lavoisier took an active, informed interest in the research of Lavoisier's assistants.[9]

For Paulze Lavoisier, an enlightened French wife, being Lavoisier's humble assistant would not have been enough. Instead, she regarded their union as a complementary partnership, one to which she could bring essential skills that he lacked. Even as a teenage bride, Paulze Lavoisier immediately identified one of Lavoisier's limitations – some of Europe's most important chemical research was being carried out in Britain, but he could not speak English. As soon as they were married, she started to learn English so that she could translate books and papers into French for him. She also engaged a Latin teacher, and begged her older brother to help her 'decline and conjugate for my own pleasure and to make me worthy of my husband'.[10]

Drawing was another talent that Paulze Lavoisier developed. She had lessons with David, who took on an unusually high number of female students. Franklin was delighted with the portrait of himself that she sent him, and she probably sketched other friends as well. More importantly for chemistry, she learned the professional drafting techniques that she needed to make accurate scale drawings of chemical apparatus – drawings that were vital for promoting Lavoisier's revolutionary chemical ideas.

She asked Lavoisier to teach her chemistry, and they also arranged private tuition for her at home. Later, she went out to lecture courses at one of Paris's private colleges. A fellow student sneered that 'she was young, but not pretty. Somewhat pedantic, she had an excessively high opinion of herself.' He accused her of stinginess, reporting that she walked home at night instead of taking a private carriage. Was she, perhaps, taking a secret detour to meet her lover, Du Pont de Nemours, before returning to her husband and his constant business meetings?[11] In any case, she became very proficient at chemistry. When the English agricultural economist Arthur Young came to tea, he was impressed by her knowledge: 'Madame Lavoisier, a lively, sensible, scientific lady, had prepared a *dejeuné Anglois* of tea and coffee; but her conversation on Mr Kirwan's Essay on Phlogiston . . . and on other subjects, which a woman of understanding, who works with her husband in his laboratory, knows how to adorn, was the best repast.'[12]

Who works with her husband in his laboratory: this was no empty flattery, but a visitor's report about real life in the Lavoisier household. Perhaps Young had been invited to attend one of Lavoisier's Saturday sessions with his students, the high point of his week. Saturday was for him 'a day of happiness', wrote Paulze Lavoisier, when 'a few enlightened friends, a few young people . . . gathered in the laboratory from the morning: it was there that we ate lunch, discussed, and created the theory that immortalised its author'.[13] Paulze Lavoisier's sepia drawings provide unique and invaluable information about Lavoisier's laboratories and research techniques. Figure 25 shows Lavoisier and his team carrying out experiments on respiration. Only two such pictures survive, and they provide unique evidence of the particular instruments he used and the laboratories where he worked.

In the centre sits the subject, dressed in an impermeable suit of rubber-coated taffeta and pumping a foot treadle. So that the amount of oxygen

Fig. 25
Marie Paulze working in the laboratory.
Marie Paulze Lavoisier, 'Experiments on the respiration
of a man carrying out work', probably 1790–1.

he consumes can be accurately measured, he breathes from a flask through a tube that is held in his mouth with putty. Paulze Lavoisier is at her usual place by the small side table, recording the observations as the experimenters call them out – many pages in Lavoisier's laboratory notebooks are in her handwriting. She was evidently a systematic, organised woman, since her other scientific tasks included filing the notes that Lavoisier scribbled on the backs of envelopes and playing cards. She has produced a Davidean portrait of a laboratory, in which the high shelf of flasks provides a stage-like setting for dramatic action. With his back to the spectator, Lavoisier theatrically directs operations, while Paulze Lavoisier is holding the same pose as David gave Lavoisier in their double portrait. By composing the picture as though she were an external spectator, she emphasises her status as a detached observer both inside and outside the laboratory.

A century elapsed before these two sketches appeared in print, but thirteen of Paulze Lavoisier's meticulously drawn plates were published in Lavoisier's 1789 *Elements of Chemistry* (Figure 26). Precise instruments were central to Lavoisier's chemical reforms, and Paulze Lavoisier's illustrations

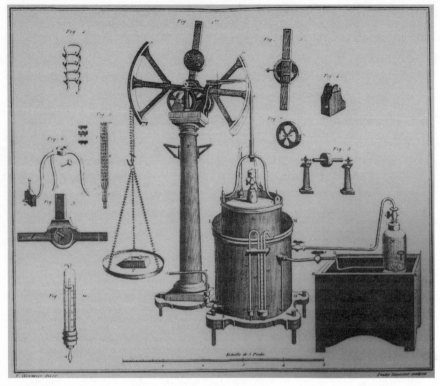

Fig. 26
One of the thirteen plates by Marie Paulze in Lavoisier's
Elements of Chemistry.
Marie Paulze Lavoisier, 1789.

made his pioneering text different in style and scope from previous ones.
Instead of the rough impressionistic drawings that characterise older books,
Paulze Lavoisier produced accurate scaled diagrams that would enable
other chemists to build their own identical instruments and so reproduce
Lavoisier's results. As Young commented on the gasometer shown in Figure
26, 'it is a splendid machine . . . too complex to describe without plates'.
Paulze Lavoisier's diagrams were a vital component of Lavoisier's book,
the propaganda piece for his new chemistry. She took great care over
them. First she painted the instruments in watercolours, and then, using
a pencil, she carefully copied these pictures on to squared grids that she
had ruled out herself. After engraving her own copper plates, she supervised several proof stages, adding corrections until she was satisfied. *Bonne*

(good) she finally wrote in approval, and signed the plates *Paulze Lavoisier* – the name she also used for business correspondence.[14]

By the time that Young arrived for tea, Paulze Lavoisier was corresponding with Europe's leading chemists, attending their lectures and commenting on the latest scientific news. As Young remarked, she could converse so knowledgeably with him about Richard Kirwan's *An Essay on Phlogiston* because she had translated it from English into French. She had worked on other British chemistry books and papers for Lavoisier's personal benefit, but because her translation of Kirwan was published, it made a vital contribution to Lavoisier's campaign of self-promotion. Kirwan, an Irish chemist, defended phlogiston, a theoretical substance that had been devised to interpret experiments on burning.

If Lavoisier wanted to revolutionise chemistry, it was essential for him to demolish the phlogistic views of Kirwan and his British allies. As part of their campaign, the couple even staged a theatrical mock-inquisition, in which Paulze Lavoisier played the part of a high priestess sacrificing phlogiston's supporters on the altar of Lavoisier's truth. For her French edition of the Kirwan book, she planned an elaborate allegorical frontispiece that would show the triumphant spirit of chemistry brandishing a flaming torch over the blindfolded, defeated figure of phlogiston. Although this was dropped, Paulze Lavoisier did provide the translation, the preface and some notes, while further criticisms were added by Lavoisier and his supporters.[15]

In any case, Paulze Lavoisier knew enough chemistry to pick holes in Kirwan's work herself. In her French version of one of his articles, which was published in the prestigious *Annales de Chimie*, she added her own critical comments. Hiding behind the anonymity of a 'Translator's note', Paulze Lavoisier formulated her denunciations – Kirwan's forgotten to mention the water that was produced; the ammonia he used was obviously impure; he seems to have made a mistake here. In voicing such comments on Lavoisier's behalf, she was, perhaps, transgressing the unwritten rules of French intellectual marriages, which were based not on equality, but on complementarity. He was the public man of reason; her role in their collaboration was to provide the skills that Lavoisier lacked – illustration, translation, formal entertainment, routine household and laboratory administration.

Twenty-three years of apparent stability and happiness: a long time, cut short only by Lavoisier's arrest and execution. Like many relationships, theirs flourished because they spent time apart as well as together. In Paris, they worked with each other in their laboratory, and enjoyed entertaining their frequent guests as well as visiting other people. They were both deeply interested in music – Paulze Lavoisier probably played the piano – and went to the opera, ballet and art exhibitions. Yet Paulze Lavoisier also had long hours by herself to study, draw and see her lover Du Pont de Nemours. Three times a year, Lavoisier spent a few weeks at his model farm in the country, but Paulze Lavoisier was unenthusiastic about the conservative neighbours, and often found excuses for retreating back to Paris and Du Pont de Nemours. But she did not always stay behind. As Lavoisier threw himself into schemes for reforming chemistry, agriculture and government, Paulze Lavoisier travelled round France with him. Their itineraries were varied and intensive – a sugar refinery in Orléans, a chemical laboratory in Dijon (home of another chemical wife, Claudine Picardet), a glove factory in Vendôme . . . Lavoisier's ideology of peaceful revolution involved hard work.

He was committed to improving the life of his peasant farmers, and he tried out new agricultural methods with the same precision as his laboratory experiments. Grain yields, milk production, craft supplies – everything increased under his methodical direction. Yet, as he constantly complained, the taxation system meant that he was earning virtually no income from the small fortune he poured into the land. The countryside was being starved, he protested, because the government provided no incentives for landowners to invest. Paulze Lavoisier shared his political idealism. At the farm, she kept production records and promoted local light industries to bring in money during the winter; together, the couple toured round factories in the hope of boosting the depressed economy. Paulze Lavoisier did her sums and made her recommendations. The profitability of a man-powered cotton mill was less than one using water power, she calculated; instead of spinning hemp, workers could earn more money by knitting Turkish hats – she suggested three a day selling at four sous each.

She also accompanied Lavoisier when he went to superintend another of his pet projects, a gunpowder factory where his reforms had more than doubled productivity in seven years. This was a sobering trip. During one round of experiments, another chemist and his sister were killed, and the Lavoisiers only escaped death because they had taken an hour off to have breakfast. They resolved to tighten the security precautions. 'Only two minutes later,' she wrote in shock, 'and six of us would have been victims.'

When a reporter for the *Journal de Paris* described this fatal explosion, he remarked disparagingly that 'discoveries of this sort are more harmful than advantageous to humanity'. Lavoisier and Paulze Lavoisier disagreed. They were both dedicated to achieving social reform through scientific research, and continued their heavy round of chemical experiments, agricultural innovations and factory visits.[16]

The accident had happened on 31 October 1788. Five years later, after a series of warnings as the Reign of Terror intensified, their lives changed far more abruptly.

4 frimaire An II: the Convention discusses tax collectors. By a show of hands, the revolutionaries agree 'that those public bloodsuckers should be arrested and that, if their accounts are not presented within a month, then the Convention must hand them over to the sword of the law'.[17] After a few days hiding with friends, Lavoisier gives himself up, and – accompanied by his father-in-law – joins the 200 inmates in the Port-Libre prison, where he remains for more than five months.

19 floréal An II: Lavoisier is searched, stripped of all his valuables, and brought before the Revolutionary Tribunal. That evening, hands tied behind his back, he stands behind his father-in-law in the place de la Révolution. Twenty-eight men are guillotined in thirty-five minutes – almost one a minute. 'It took them only an instant to cut off that head,' commented one of Lavoisier's scientific colleagues, 'but it is unlikely that a hundred years will suffice to reproduce a similar one.'[18]

The raw facts: undisputed, so straightforward to narrate. Motivations and allegiances remain less clear. Despite his radical ideals, Lavoisier was executed because his wealth made him a symbol of political oppression. Could he perhaps have saved himself? After all, many other wealthy men fled from Paris or bribed their way out of prison. And what was Paulze

Lavoisier doing during the Reign of Terror? Surely she could have persuaded one of their influential friends to rescue him?

She visited Lavoisier regularly while he was in prison, liberally tipping her way round the warders to ensure that her husband and her father had enough food, clothes and firewood. She also helped Lavoisier to plan his defence, sending letters back and forth between the prisoners and their supporters on the outside. The furniture in their Paris flat was taken over by the revolutionaries, and thieves stole most of the livestock from their farm. They both knew that he was unlikely to survive. 'My dear one,' wrote Lavoisier, 'you are . . . exhausting yourself both physically and emotionally, and, alas, I cannot share your burden.' Be careful of your health, he continued; 'my task is accomplished. But you, on the other hand, still have a long life ahead of you.'[19]

Both Paulze Lavoisier and Lavoisier seem to have spurned the possibilities of negotiating their way out of the situation. When a friend engineered an interview for her with Antoine Dupin, the official responsible for drawing up the charge sheet against Lavoisier, she stormed into his office, announcing that she refused to humble herself by begging for pity from a Jacobin. She sought justice not mercy, she declared: a noble attitude, but diplomacy or bribery might have been more appropriate.

Her life became still harsher after Lavoisier was guillotined. She wandered round the empty apartment. Furniture, instruments, ornaments – all gone, all confiscated. The only book remaining on the shelves was the catalogue of 560 vanished volumes, while the pale blocks on the panelling reminded her of the pictures that had disappeared. She knew that her own arrest was imminent, yet constantly put off taking refuge in the countryside with Du Pont de Nemours. Perhaps she was worried about becoming involved once again with a far older man. Or perhaps his attractiveness waned once he was available, no longer a forbidden fantasy. And there was also a financial complication: Du Pont de Nemours had borrowed money from Lavoisier to set up his printing plant, but was no longer able to keep up the debt repayments. Paulze Lavoisier procrastinated too long. On 22 messidor (10 July 1794) Du Pont de Nemours received a coded letter from his son: 'Citizen Lavo,' he learned, 'has met with a very great tribulation . . . we hope that she will be happily extricated in a few days.'[20] A few? She spent sixty-five days in prison, an ordeal she could have avoided by taking a less compromising attitude towards her accusers, as well as her friends.

Paulze Lavoisier was exonerated from suspicion of political activism by her scientific dedication. 'It can be assumed,' the surveillance committee reported, 'that collaborating daily with her husband in his work, she was involved only with what related to their domestic occupations.'[21] Paulze Lavoisier spent the next couple of years plotting retribution on Dupin and his allies. Du Pont de Nemours obligingly printed the brochure she wrote accusing Dupin of being a murderer, and he was eventually tried and demoted. Then she set about systematically searching for her confiscated possessions, pestering government departments for their return. Much of it came back – paintings, maps, furniture, even two bottles of mercury and twelve notebooks (six used, the inventory noted). After the farm and the Paris apartment were restored to her in April 1796, her financial security was assured.

As well as recovering her husband's property, Paulze Lavoisier also wanted to rebuild his reputation. His former colleagues, worried about accusations of failing to protect him, were converting Lavoisier into a national hero. Modern chemistry was, they declared, a French creation, an enduring contribution to knowledge whose permanence was very different from the unseemly haste with which the former regime had removed his head. Lavoisier became doubly celebrated as a champion of experimental science and a martyr to Jacobin irrationality. A theatrical funeral ceremony attracted 3,000 visitors, but Paulze Lavoisier was not among them – she never forgave their friends for being unable (or unwilling?) to save her husband.

For her own tribute to his memory, she decided to publish his last manuscripts. While he was in prison, Lavoisier had been working on an edited collection of papers, but although Du Pont de Nemours started printing these *Memoirs of Chemistry*, he never completed the job, so Paulze Lavoisier resurrected the abandoned project. (The Du Pont de Nemours family emigrated to America and founded the successful chemical industry, even toying with the idea of naming their Delaware gunpowder factory Lavoisier's Mills.)

Years slipped by as Paulze Lavoisier tried to persuade one of Lavoisier's collaborators (the man inside the rubberised suit in Figure 25) to write a preface condemning Lavoisier's persecutors. Since he refused, she provided her own short, unsigned preface, which was designed to reinforce Lavoisier's arrogant boast that the new chemistry had been devised by him alone: 'elle est *la mienne*,' he had written – 'it is *mine*'. She distributed copies to Paris's leading men of science, and accolades poured in for France's scientific icon. Even one of Lavoisier's former opponents commented that 'The name of Lavoisier is a greater recommendation for these new memoirs than all the praise that we could give them.'[22]

The *Memoirs* finally appeared in 1805. No coincidence, perhaps, that this was the year she married her second husband, the American Benjamin Thompson, an expert on the physics of heat and light. In Figure 29, he is the man with a bulbous nose standing by the door to the right of the stage. Like Lavoisier, Thompson was committed to finding practical applications for his scientific research; as a reward for his work to help the poor in Munich, he had been elevated to Count Rumford. Ostensibly the ideal bridegroom, he courted Paulze Lavoisier for four years, as they visited each other's homes and toured round Europe together. At last he succeeded in defeating other suitors. 'She appears to be most sincerely attached to me,' Rumford wrote to his worried daughter, 'and I esteem and love her very much.'[23]

But only months after the wedding this romantic relationship disintegrated, culminating in an explosive row when he locked out some of her friends and she took revenge by pouring boiling water over his carefully cultivated roses. What went wrong? Rumford called Paulze Lavoisier 'a female dragon', a 'tyrannical, avaricious, unfeeling woman'; his last six months with her were, he moaned, 'a purgatory sufficiently painful to do away the sins of a thousand years'.[24] Her version of events has not survived, but was undoubtedly different. The evidence that does exist suggests that much of the blame lay with him. Even Paulze Lavoisier's step-daughter – not the easiest of relationships – found her charming and easy to get on with. In contrast, several contemporaries criticised Rumford for being an arrogant, condescending and tactless man. Although he dispensed help to the needy, they sniped, he obviously despised them.

Money was definitely a source of strife. She owned a vast fortune, whereas he was struggling to survive on his Bavarian pension. Rumford eventually walked away with a divorce settlement equivalent to well over a million dollars – a gold-digger, sneered the satirists. Paulze Lavoisier's intellectual prowess also came between them. Instead of the French concept of sexual complementarity, Rumford held an Anglo-American view of how women should behave. At first he boasted about his new wife's cleverness and the magnificent *soirées* that she threw, but he rapidly became angered by her independence and her extravagance with her own money.

Once disentangled from Rumford's oppressive regime, Paulze Lavoisier returned to Paris and a further twenty-seven years of scientific entertainment and discussion. She entertained two or three times a week: a small select dinner on Mondays, open house on Tuesdays, and frequent Friday evenings of music and conversation. Men of science, artists and poets, foreign visitors – all were invited to mingle at her *salon*, which was renowned not only for its diversity, but also for its freedom of thought.

'You have to have lived under the vacuum pump to appreciate the luxury of breathing,' reminisced one admirer. Well into her seventies, Paulze Lavoisier still supervised the food and the dancing, but also cross-examined her guests about their work. Except for Lavoisier's absence, it was like re-entering the eighteenth-century world of polite conversation, almost as if the Revolution had never happened. But Lavoisier never fully left her – David's double portrait (Figure 24) dominated the drawing-room where, curled up on a love-seat, she welcomed her visitors with a disarming mixture of rudeness and politeness.[25]

Under Science's Banner

It cannot, I think, be truly asserted, that the intellectual powers know no differ-ence of sex . . . Male genius fetches its treasures from the depths of science, and the accumulated wisdom of ages: the female finds her's in the lighter regions of fancy and the passing knowledge of the day . . . Dividing subjects of thought into abstruse, serious, *and* light, *I consider only the former and the latter as peculiarly appropriated by wither sex; the center is common to both: it is the key-note uniting two chords, equally useful and necessary to both.*

Laetitia Hawkins,
Letters on the Female Mind, Its Powers and Pursuits, 1793

'Much attention has lately been paid to the education of the female sex,' spluttered a fictional gentleman in 1795. Counselling the new father of a baby girl, he poured scorn on the idea that women could ever learn to do anything useful. Although he did admit that in 'poetry, plays, and romances, in the art of imposing upon the understanding by means of the imagination, they have excelled', he made it clear that factual matters were beyond their grasp: 'I have never heard of any female proficients in science – few have pretended to science till within these few years.' Imagination and fact, literature and science, women and men – for him, these were polar opposites.[1]

The actual author of these words was Maria Edgeworth. As part of her campaign to improve women's education, she composed an imaginary yet disturbingly realistic set of letters based on the condescending views of Thomas Day, a friend of her father Richard Edgeworth. When he was only twenty-one, Day had devised his own educational experiment. He wanted to mould a young girl so that she would fit his taste for a bride 'simple as a mountain girl . . . intrepid as the Spartan wives and Roman heroines'. Selecting two orphans – one blonde, one brunette – Day carefully followed the prescriptions of the Swiss educational philosopher Jean-Jacques Rousseau for rearing docile young girls, but was dismayed to discover that his pupils had minds of their own. Although he soon apprenticed the more rebellious one to a milliner, he persevered with the other, teaching her to endure fear and pain by firing pistols through her skirts and dropping melted sealing-wax on her bare arms. Unsurprisingly, she too was unenthusiastic about her rescuer, and ended up marrying one of his friends. Strangely, Day later published a best-selling educational book for children.[2]

Day was horrified when he discovered that Richard Edgeworth was encouraging his own daughter Maria to write about philosophy and politics. However, his advice went unheeded, and Richard and Maria Edgeworth collaborated to write books encouraging children to study and think while they played. Rational entertainment – combining learning with pleasure – was the educational fashion of the period. Girls as well as boys were learning more about science, but educators disagreed on vital issues. Should girls really learn to think for themselves, or should they continue to believe themselves governed by their emotions rather than their minds? Was it possible to reconcile intellectual independence with conventional obedience? Would fostering women's education lead to unhappy marriages and the breakdown of social stability?

Not only women were interested in reforming female education. Day and Richard Edgeworth both belonged to the Lunar Society, an informal group of fourteen men who met every month on the Monday nearest the full moon, when the country roads would be well illuminated for travelling. This rational explanation of their name did not, of course, prevent them from being known as the Lunaticks, and they exuberantly tried out many ingenious yet impracticable inventions. But they also engaged in serious discussions of scientific and technological ideas that did much to promote early industrialisation. Unfettered by aristocratic conservatism, these self-made men made the Midlands rather than London the location of change for Britain's future.[3]

Several of the Lunar men were involved in teaching women. William Withering, a medical botanist, taught female students through correspondence. 'Natural Philosophy is my only *real* Entertainment,' enthused his pupil Catherine Wright as she struggled to find time for squeezing in experiments between her household duties. Despite teaching herself chemistry and inventing a therapeutic water bath, she was embarrassed about her own ignorance. 'Excuse my follies,' she begged, 'very few of our Sex Ever Attain to the Learning of a School Boy . . . I have constantly regretted my too confined Education.' Withering repeatedly tried to discipline her imagination, dictatorially insisting that she follow his instructions rather than her own inclinations: 'I am terefied to Death while I risk my Ideas before you,' she wrote.[4]

Perhaps the most eminent member of the Lunar Society was Erasmus Darwin, who had sheltered Day's would-be bride from excessive educational experiments. Darwin bought a large house for his illegitimate daughters to run as a girls' boarding school – an enlightened act, even though he did promptly publish a pamphlet telling them how to do it. He recommended that women should be taught science, and endorsed Maria Edgeworth's work (as well as plugging his own books). Nevertheless, his advice carried a strong paternalistic edge. Armed with the latest scientific knowledge, he argued, women would be able to undertake more interesting conversations with men.[5]

Now eclipsed by his far more famous grandson Charles, at the end of the eighteenth century Erasmus Darwin was renowned throughout England not only for his educational innovations, but also for his medical expertise and radical politics. He had hesitated before publishing what turned out to be his most famous book, *The Botanic Garden*, two book-length allegorical poems explaining not only modern botany, but also many other innovations in science and technology. For one thing, Darwin favoured frankness, and knew that he would be accused of corrupting young women by using sexual terminology to describe how plants should be classified. Just as significantly, he worried about damaging his medical reputation by appearing in print as a poet. Scientific poetry had sold well throughout the century, but now new disciplinary boundaries were being set up. Men of science were distinguishing themselves from men of letters by adopting dry, sober styles of writing. Should Darwin dare to bring together facts and imagination, science and literature? Or would he be mocked for trying to reconcile incompatibles?

In the end, he wrote a preface justifying his decision. 'The general design of the following sheets,' he wrote, 'is to inlist Imagination under the banner of Science.' His anxiety was unnecessary. *The Botanic Garden* was an instantaneous success amongst literary authors as well as his scientific colleagues at the Lunar Society – 'the most delicious poem upon earth,' raved the gothic enthusiast Horace Walpole. Packed with mythological references, ornate verses and whimsical descriptions, *The Botanic Garden* is not a work that appeals to modern tastes, but at the time it was enormously popular. Poetry was well-recognised as a didactic medium, and hefty technical footnotes discussed the very latest scientific theories and technological inventions. Darwin's *Botanic Garden* became an important educational text.[6] This scientific epic influenced the major Romantic poets, including Samuel Taylor Coleridge and Percy Bysshe Shelley.

Darwin's self-exoneration – *to inlist Imagination under the banner of Science* – became a slogan that appealed to many writers. Women were traditionally credited with possessing imagination, but now some of them wanted to learn and write about science. In her calls for educational reform, Maria Edgeworth cited Darwin's rallying-call (although she did produce her own version, mis-remembering that 'Science has of late "*been enlisted under the banners of imagination*"'.) Many women, she went on, are interested in botany, but now they are turning to chemistry, formerly a male preserve. Even so, Edgeworth did not suggest that women should be involved in the excitement of research. Instead, she promoted chemistry as a safe, methodical activity, making it sound like a glorified form of cooking. 'Chemistry,' she commented in her double-edged recommendation, ' is a science well suited to the talents and situation of women . . . it demands no bodily strength; it can be pursued in retirement; it applies immediately to useful and domestic purposes . . . there is no danger of inflaming the imagination, because the mind is intent upon realities, the knowledge that is acquired is exact, and the pleasure of the pursuit is a sufficient reward for the labour.'[7]

Darwin's most famous female fan was Mary Shelley, author of *Frankenstein*, now often acclaimed as the world's greatest work of science fiction. Darwin appears in the very first sentence of *Frankenstein*'s preface, which starts: 'The event on which this fiction is founded has been supposed, by Dr Darwin . . . as not of impossible occurrence.'[8] Shelley had been particularly impressed by Darwin's account of *vorticellæ*, microscopic creatures which, after being dried for several months, seemed to be miraculously restored to life when soaked in water. Like Edgeworth, she got the details slightly wrong – *vermicelli*, she called them – but she did follow Darwin's injunction to integrate science and imagination. Excluded from real-life chemical laboratories because that would, as Edgeworth put it, court 'the danger of inflaming the imagination,' Shelley produced her trenchant critique of scientific research.

In the early seventeenth century, Francis Bacon had prescribed a new approach to learning, one based on experiments and investigation rather than on books and tradition. Bacon became an ideological figurehead whose shadow stretched forward over the centuries. He was renowned for envisaging a utopian research community where men would examine, control and exploit nature – a nature that remained feminine. 'For knowledge itself is power,' he had preached, and Mary Shelley's contemporaries still quoted his words as they put his ideals into practice.

Shelley lived halfway between Bacon and us, at a time when modern science was becoming established and women were starting to campaign for equality. Neither Bacon nor Shelley practised science themselves, yet through their writing both of them had an enormous impact on how we think about scientific research and women's roles within it. Another man-woman pair, they make appropriate book-ends for this collection of scientific partnerships which started with Francis Bacon and Lady Philosophy, and ends with Mary Shelley and Victor Frankenstein.

Chapter 10

Priscilla Wakefield/Carl Linnaeus

A consciousness of our defects is the first step towards improvement; a young lady of your age is not expected to be deeply skilled in philosophy; much less to display her knowledge, should she possess a small share; but a general acquaintance with the uses of the most common philosophical instruments is not only ornamental, but also a very useful accomplishment, and should form part of every liberal education.
Priscilla Wakefield, *Mental Improvement*, 1794-7

'Oh! that we had a book of botany,' sighed Dorothy Wordsworth. A few months later, she and her brother William had acquired a copy of the standard manual of plant classification, William Withering's *Botanical Arrangement*. Unlike other sciences, botany was deemed suitable for women as well as for men, and in Britain boasted its own female patron, Queen Charlotte, who regularly retreated with her daughters to a small green room where she could 'botanise' in peace.[1]

Flowers and femininity conventionally belonged together. 'What delightful entertainment it must be to the fair sex,' Joseph Addison had written decades earlier (sarcastically, one hopes), 'to pass their hours in imitating fruits and flowers, and transplanting all the beauties of nature into their own dress . . . This is, methinks, the most proper way wherein a lady can show a fine Genius.' Many leisured ladies whiled away the hours embroidering and painting flowers, and at the end of the eighteenth century, botany boomed when authors realised that there was an untapped market for simple instruction books. Deliberately aiming at affluent women and young girls, educators recommended botanical study as a gentle and genteel way for them to improve their scientific knowledge and gain some healthy exercise out in the fresh air.[2]

But although women were encouraged to take up botany, this was on different terms from men. The Wordsworths probably knew that Withering owed his own reputation as a heart expert to the traditional remedy revealed to him by an elderly village woman. Rather like the male-midwife (see Figure 6), Withering had taken over the knowledge of female herbalists and given it a new scientific authority: instead of a foxglove recipe, he prescribed precise doses of digitalis. In contrast, with a few notable exceptions, women were restricted to collecting, drawing and writing about plants rather than making new scholarly discoveries.

Men made recommendations that seemed to endorse female involvement, but in fact reinforced their exclusion from intellectual work. One gentlemanly journal correspondent insisted that women's delicacy made them particularly good at growing tender imported plants. In his opinion, it seemed clear that botany was 'an elegant *home* amusement . . . more dextrously performed by the pliant fingers of women, than by the clumsy paws of men'. And, he continued, women were better at coping with 'the regulation of the green-house sashes' – the servants were not to be trusted, the gentlemen were too busy, and so the obvious candidates for this onerous task were women.[3]

Converting botany into a scientific discipline involved making it systematic. In one of the many botanical poems of this period, Flora – the goddess of flowers – explains to Jupiter why spring has arrived abnormally late. The poem's anonymous author (probably Erasmus Darwin's friend Anna Seward) was writing tongue-in-cheek, but sarcasm works best at the painful edge close to reality. Although deciding how to classify flowers might seem an innocuous topic, it aroused intense feelings throughout the eighteenth century. Botanists were searching for God's natural blueprint, yet the order they imposed on the plant world mirrored the social patterns of the human world.

Flora reports that Britain's flower beds are becoming hopelessly entangled because the plants have not been classified properly:

> Vegetation of course was o'errun with disorder
> From the wood & the wall to the bank & the border.

Scientific organisation in the countryside, argues Flora, is as important as the traditional hierarchies that maintained the stability of the British nation:

> Here rank & high titles, says she have no merit
> And my Weeds are brought up in a leveling spirit
> You vagabond Fungus what else cou'd provoke
> To tread on the toes of his highness the oak.

But, Flora continues reassuringly:

> Rejoice then my Children the hour is at hand
> When Botanical knowledge shall govern the land.[4]

Flora's saviour, her hero who brought order to the botanical kingdom, was Carl Linnaeus. Figure 27 shows the brilliantly coloured engraving that formed the frontispiece of a large expensive book about flowers called *The Temple of Flora*. Linnaeus is literally up on a pedestal, being worshipped for his three contributions to human welfare – medicine, agriculture and botany. On his left stands Aesculapius, the god of healing (with whom Withering would probably have liked to identify himself); to the right, holding a scythe and honouring Linnaeus with a laurel wreath of fame, is Ceres, the goddess of agriculture, the mainstay of Britain's wealth in this pre-industrial period; and at the front kneels alluring Flora – the emblem of botany, that 'elegant pursuit for ladies'. The startling red cloaks of Aesculapius and Ceres contrast with Flora's white robe of chastity, while the bright chains of flowers in primary colours may have been arranged to deliver a coded message in the 'language of flowers' – the traditional symbolism, most famously expressed by Ophelia in *Hamlet*, which accords different sentiments to different flowers.[5]

Carl Linnaeus is still a scientific hero, especially in his native country of Sweden where Linnaean cakes and mineral water (appropriate souvenirs for a celebrated teetotaller) boost the national tourist revenue. How did Linnaeus, an old-fashioned pastor from a small provincial town, become world-famous? In conventional, romanticised versions of science's history, great men succeed because they have great ideas . . .

Fig. 27
Aesculapius, Flora, Ceres and Cupid honouring the bust of Linnaeus.
From Robert Thornton's *Temple of Flora* (1806).
Engraved by Caldwell after John Russell and John Opie.

The eighteenth century is often called the Age of Classification, and Carl Linnaeus was the classifier *par excellence*: as his colleagues quipped, 'God created and Linnaeus organised.'[6] Previous botanists had proposed many different schemes for grouping plants, based on criteria such as the colours of flowers or the shapes of leaves, but none of them worked satisfactorily. In 1732, Linnaeus resolved this dilemma by introducing a startlingly new yet simple system – he classified plants according to the number and arrangement of their reproductive organs, the male stamens and the female pistils in the flowers. By counting the stamens, he divided plants into twenty-four classes, and then divided each of these into sub-groups called orders, defined by the pistils. Using additional character-istics, he made finer and finer divisions, which enabled him to identify each plant by a double Latin name: *Citrus limon*, for instance, is a lemon tree, while *Citrus aurantium* is an orange tree. He later extended this bino-mial scheme to cover animals, which is why human beings are known as *homo sapiens* – wise man.

As a young man, Linnaeus travelled to Lapland, bravely venturing northwards far beyond the Arctic circle. There he discovered a small plant which, following his own binomial system, he named after himself – *Linnaea borealis*, Linnaea of the north. Its frail white flower became a Linnaean emblem, painted on to china and prominently displayed in all his portraits. After his return to southern civilised Sweden, Linnaeus spent most of his life at the University of Uppsala, where he laid out his botanical garden like a modern-day Garden of Eden. By arranging foreign and native plants according to his new classifi-cation scheme, Linnaeus aimed to mirror God's original design for the Earth.

The great virtue of Linnaeus's botanical system was its simplicity. It was easy to understand and explain, and because it relied on counting, personal judgements were eliminated. Non-experts – even women – could be trusted to follow the instructions. Linnaeus enthusiastically dissemi-nated his doctrine, marshalling his ideas as rigorously as his flowers into twelve chapters and 365 aphorisms – a Linnaean thought for every day of the year, like a Lutheran almanac. Linnaeus was committed to making his classification system useful for non-university-trained people, including women. He dreamed of making his own 'Language of Flowers' universally accessible, boasting that 'it is as quick to read Nature as any other Book; yes, even for Women themselves.'[7]

Linnaeus wanted to reform Sweden's economy by producing locally many of the exotic crops being imported from abroad. Although he himself remained in Uppsala, he sent out his students to gather plants from around the world and bring them back to his garden, where he initiated ambitious projects to produce home-grown substitutes for coffee, tea, spices and other luxuries. As in other European countries, wealthy women were demanding these fashionable yet expensive imports.

This was a two-way trade in goods and ideas – Linnaeus's recruits, whom he called his 'apostles', also had the task of spreading the Linnaean gospel to botanists in other countries, including Britain. These Linnaean disciples accompanied explorers on their voyages to the Pacific and other remote regions, showing them how to describe and classify the astonishing new plants they discovered. One of the first institutions to adopt Linnaean classification was the British Museum, where the new Swedish system was used for displaying the exhibits and laying out the gardens in satisfyingly ordered arrays. Linnaeus's taxonomy elbowed out the opposition and became adopted as the international standard.

So is there anything unsatisfactory with that account of Linnaeus's importance? Does it neatly take account of all the major facts, or could it be retold in a different way? One shortcoming is that it diminishes the importance of Linnaeus's own zeal in self-promotion. Radical as his system was, it might well not have been so successful had he not advertised it so relentlessly. And as well as producing propaganda for his classification scheme, he carefully adjusted some details of his own life story. He presented himself as an intrepid Arctic explorer, even though he had secretly stashed away some helpful older guidebooks in his luggage; on his return, he falsified the reports he submitted to the Academy of Sciences to make it look as though he had travelled far further than he actually managed. His favourite portrait, especially designed to win over potential patrons, shows him dressed in a supposedly authentic Lapp costume, although in reality it was a motley assortment of tourist souvenirs.

More significantly for the topic of women in science, conventional versions of Enlightenment botanic history are misleading because they imply that Europe's botanic experts were rapidly converted to Linnaean classification, and that it soon became the basis of a new science for women. Certainly, botany was big business at the end of the eighteenth century. A large book called *The Temple of Flora* (Figure 27) was conceived as a commercial venture, in which – similarly to projects on Shakespeare, Milton and the Bible – famous artists were commissioned to produce ornate paintings of flower arrangements. These were exhibited in public galleries as well as being engraved for this handsome folio volume. However, even with the help of a last-ditch botanical lottery, the ambitious scheme turned out to be a financial flop. This failure suggests that Linnaeus's rise to fame was more complicated than it might seem. The two small cherubs in the picture provide a clue.

Cushioned on his billowy cloud, the zephyr with butterfly wings symbolises spring, the most rewarding season for studying botany. Appropriately, he is decorating Linnaeus with flowers. Because Linnaean classification depended on flower parts, botanists were at this time focusing their attention primarily on flowers rather than on leaves and stems. Unlike most objects of scientific study, flowers were strongly linked with women. Still worse, as far as upright eighteenth-century Englishmen were concerned, Linnaeus's scheme was based on sexual reproduction – and flowers were traditionally metaphors for sexual organs.

Linnaean classification was enormously controversial because it was based on sex. The winged boy in front of Linnaeus's column is no innocent observer, but Cupid with his arrow of love. Words like 'defloration' indicate the long ancestry of the association between human anatomy and flower parts, one that still resonates in some flower pictures, such as those by the American artist Georgia O'Keeffe. From the end of the seventeenth century, when the sexual nature of plants had first been established, the book trade provided many pornographic botanical poems, and in *The Temple of Flora* several of the magnificent coloured plates showing exotic flowers are highly eroticised.[8]

For Linnaeus, the upright son of a Lutheran parson, sex meant marriage, and his system transferred orthodox human relationships on to the plant world. Linnaeus organised his flowers in the same way as he ran his life, with males at the top of the hierarchy. Just as men dominated European society, so too Linnaeus's major classification into orders was determined by the number of male stamens. Linnaeus may have been promoting strait-laced family values, but even so, the vocabulary he used to describe his flowers seems extraordinary. 'The flowers' leaves,' he wrote, 'serve as bridal beds which the Creator has so gloriously arranged, adorned with such noble bed curtains, and perfumed with so many soft scents that the bridegroom with his bride might there celebrate their nuptials.'[9] Bride and groom might be tolerable analogies for plants with one pistil and one stamen, but comparisons became more problematic in the higher classes and orders, when multiple partners were sharing the nuptial chamber.

Far from being an overnight success as traditional accounts would have it, Linnaean classification was only slowly adopted. Although his scheme was well known by the middle of the eighteenth century, it was still controversial fifty years later. In Britain, plant sexuality was a major obstacle. English moralists were appalled by what they called this 'florid' style (although they evidently liked puns). However simple it might be to understand, undiluted Linnaean sexuality was, they judged, far 'too smutty for British ears' – 'nothing could equal the gross prurience of Linnaeus's mind . . . enough to shock female modesty'.[10] In 1798, the Reverend Richard Polwhele remarked critically that 'Botany has lately become a fashionable pursuit with the ladies. But how the sexual system of plants can accord with female modesty I am not able to comprehend.' He had even, he gasped, 'several times, seen boys and girls botanising together'.[11]

Advocates of female education, men like Withering and Darwin, were caught in a dilemma. Botany seemed an ideal science for women to study, yet Linnaeus's sexually charged language was totally unsuitable. Withering resolved the problem by removing references to plant sexuality when he translated Linnaeus from the original Latin into English. Unfortunately, although this editing did sanitise botany, it also made the structure of Linnaeus's classification scheme incomprehensible. Darwin adamantly disagreed with Withering. First, he produced his own scholarly translation, and then – prompted by Anna Seward – he set out to compose Linnaean poetry. Following his own dictum to enlist imagination under the banner of science, Darwin published *The Loves of the Plants* (later part of *The Botanic Garden*), a long poem celebrating floral sexuality, which would, he hoped, appeal to 'ladies and other unemploy'd scholars' (an interesting bracketing!).[12]

Linnaeus had mapped social hierarchies onto the plant world by making the number of male stamens his primary classification criterion.

As so often happens in science, Darwin helped to reverse the process, so that Linnaeus's anthropomorphic system came to prescribe how people should be classified. If male superiority structures nature, the logic goes, then that is the way we should organise society – conveniently overlooking how the concept has been borrowed from the human world in the first place.

Darwin versified flower structures in terms of human relationships: he eroticised the system that Linnaeus had sexualised. Writing in rhyming couplets, Darwin bracketed together botany and mythology to portray a romanticised vision of an erotic paradise. Like a literary Capability Brown, he created a magnificently profuse yet orderly botanical garden that appeared deceptively natural yet also reflected the desires of Georgian gentlemen. All sorts of female stereotypes appear in *The Loves of the Plants* – the virtuous virgin, the timorous beauty, the laughing belle, the dangerous siren.

For more than 1,700 lines, Darwin kept up a style that now seems tedious, but then was hugely popular. For example, this is how he describes the plant *Collinsonia*, which has two male stamens and one female pistil:

> *Two* brother swains, of COLLIN's gentle name,
> The same their features, and their forms the same,
> With rival love for fair COLLINIA sigh,
> Knit the dark brow, and roll the unsteady eye.
> With sweet concern the pitying beauty mourns,
> And sooths with smiles the jealous pair by turns.[13]

A contributor to the *Encyclopædia Britannica* had found this second class of plants particularly obnoxious. 'A man would not naturally expect to meet with disgusting strokes of obscenity in a system of botany,' he expostulated. 'But . . . obscenity is the very basis of the Linnean system.'[14] The reactionary Reverend Richard Polwhele agreed, although his own choice of vocabulary doubtless helped boost his sales. Polwhele evoked the familiar figure of Eve as he luridly portrayed the consequences of allowing women to practise botany and so lure men to their doom:

I shudder at the new unpictur'd scene,
Where unsex'd woman vaunts the imperious mien . . .
With bliss botanic as their bosoms heave,
Still pluck forbidden fruit, with mother Eve . . . [15]

Writing with Mary Wollstonecraft as his specific target, Polywhele parodied Darwin's verses on *Collinsonia* by envisaging her simultaneously tending two lovers, a scenario that conformed to the current gossip about her affairs. Nevertheless, with two men and one woman, Darwin was still on relatively safe territory, and he placed the Collins trio in an idealised, harmonious realm populated by compliant nymphs and laughing shepherdesses. However, his tone shifted when he reached higher orders, which had an even larger male-to-female ratio. In these socially unacceptable situations, his women were no longer submissive adoring maidens, but became either seductive promiscuous flirts or powerful arrogant sovereigns.

Other types of women were notable by their absence. Despite Darwin's support for his daughters and the cause of female education, despite his friendship with Maria Edgeworth and other writers, there were no intellectual women in his botanic garden – not even any artists, poets or novelists, let alone scientific women. Erasmus Darwin may have professed progressive attitudes, yet his female stereotypes were, like those of the prudish Reverend Polwhele, more like Eve than Minerva.[16]

In 1796, two important books about Linnaean botany appeared. One was a new edition of Withering's *Botanical Arrangement*, the version bought by Dorothy and William Wordsworth. By then, Withering had capitulated and adopted the sexual terminology that Darwin insisted on using. The other was Priscilla Wakefield's *Introduction to Botany* – a book on botanical science written by a woman for women. This slim volume, illustrated by her own detailed drawings, became the standard introductory text for the next fifty years, and was so successful that it ran through nineteen editions in England, America and France.

Being a best-selling scientific author is, of course, an achievement, but Priscilla Wakefield (1751–1832) is also interesting for other reasons. Her *Introduction to Botany* is ostensibly about Linnaean classification, but it is also about social organisation. Railing against the high prices and Latin terms that prevented ordinary women from studying botany, Wakefield was more interested in reforming society and female education than in modernising botanic science. Her book on botany was the first major example of a new wave of books that taught girls about science. That might seem heartening, but this new educational fashion did little to help women participate in scientific research. Furthermore, by steering women towards botany, Wakefield's book reinforced their unsuitability for other types of science.

Like many authors, Wakefield used her writing as a form of intellectual therapy to distract her from life. '*Employment*,' she observed, is 'a powerful remedy for *uneasiness* tho' there are many causes that simply depress me.'[17] She worked assiduously, squeezing in sessions between the frequent demands of sick relatives, children and household chores. Snatched time alone was rare and precious – 'such a day now & then is a feast,' she confided to her diary. With its lines neatly ruled out by hand, Wakefield's journal provides evidence of her thoughts as well as her activities (and also includes a one-inch margin for that English preoccupation, the weather).

An energetic but depressed and overburdened woman, Wakefield found writing difficult. But she needed the money – 'necessity obliges me to *write*,' she exhorted herself.[18] Weighed down by worries about the family's finances, she started publishing small books for children when she was already forty-three, well into old age by the conventions of the time. She had sensibly married a wealthy man, but as he made one unwise business decision after another, his inherited fortune slipped away, and two of their sons were also floundering. Wakefield was a resourceful woman. When their house caught fire, her husband apparently stood by crying while she organised a chain of people to pass buckets of water.

Along with other female authors of this period, Wakefield was helping to establish that writing educational texts could provide a viable career. She turned to natural history not out of passion for the subject, but because she was desperate for money and had sized up a gap in the market. Studying by herself from encyclopaedias and standard science textbooks, Wakefield constantly edited and rewrote, and struck hard bargains with her publishers. Approaching her self-appointed task with business-like determination, she refused to be deterred by unfamiliar subjects. As well as botany, she produced simplified accounts of insects, fish and other topics in natural history, and took her young readers on imaginary tours throughout the world so that they could absorb the principles of geography. By the time she died, Wakefield had written seventeen books, many of them appearing in multiple editions and even translated into foreign languages.

The member of an eminent Quaker family, Wakefield lived in Tottenham, which was then a prosperous middle-class village outside London. Like many other affluent Quakers, Wakefield and her family devoted themselves to good causes. Her niece grew up to be Elizabeth Fry, the prison reformer now shown on Britain's five pound notes. However, the life of one of her sons was more chequered: he eloped with a teenage heiress, and then tricked another rich girl into marriage at Gretna Green, for which he spent three years in Newgate prison.

Because Quakers were not allowed to study at university, they were used to running their own schools and academies. Wakefield participated in ambitious schemes of educational reform, drumming up support from influential patrons. Before financial ruin threatened her family, she established a maternity hospital for impoverished mothers and was involved in setting up new schools for girls. Evidently possessing sounder business acumen than her husband, she set up pension schemes and a Penny Bank to encourage poor people to earn interest from their money. These charitable activities gained her one of those heroic 'firsts' so beloved by some scholars – in banking histories, Wakefield's 'Frugality Bank' features as the first savings bank.

Wakefield wanted better conditions for women, but – like many of her female contemporaries – she insisted that she was not 'one of those bold projectors who are desirous of overturning the present system of society, by placing women on an equality with men'.[19] Motherhood was important for her, and she stressed the value of family stability. Grateful to her own 'most excellent of mothers', she taught her five children as well as several of her badly behaved grandchildren, using her favourite, Kitty, as a guinea pig for her books.[20]

Women as well as men promoted ideals of female domesticity. Women needed to be better educated, the argument ran, not so much for their own benefit but to nurture future generations of male scholars and provide stimulating fireside and dinner-table conversation for their men. Educational reformers emphasised that families should carry out joint activities, and mothers were being encouraged to teach their children at home. Even some of the masters at England's public schools were recommending that small boys should first be taught by women. Enterprising women set up small schools for girls and catered for the growing market in female education by writing books specifically targeted at children and young women.[21]

This maternal approach to teaching is illustrated by the frontispiece of another early science text for girls (Figure 28) – Margaret Bryan's book on astronomy, which came out the year after Wakefield's *Introduction to Botany*. Apart from her learning and her appearance, virtually nothing is known about Bryan, although her books were well regarded. Initially, even the notion that a woman might write a textbook on such a physical subject seemed strange. When the original painting for this frontispiece was exhibited at the Royal Academy, the catalogue mistakenly

Fig. 28
Mrs Bryan and children.
Frontispiece of Margaret
Bryan, *A Compendious System
of Astronomy* (1797).
Engraving by W. Nutter
from a 1797 miniature by
Samuel Shelley.

reported that it had been designed for a book by a *Mr* Bryan. Nevertheless, for more than twenty years Mrs Bryan ran a succession of girls' schools, and was encouraged to publish by a distinguished (male) professor of mathematics.

As the picture suggests, although Margaret Bryan taught scientific subjects that had conventionally been restricted to boys, she encouraged a gentle, maternal approach towards her female pupils. Scientific instruments and quill pens normally signified the masculine world of scholarship, yet this scene radiates a soft domestic harmony. The two girls nestle close to their teacher, who seems more like a friendly guide than a stern instructress. Bryan ended each chapter with a moral homily about good behaviour and the benefits of studying natural philosophy. Sciences like astronomy and physics were, she insisted, perfectly compatible with the traditional feminine graces of virtue, charm and benevolence. Her sanctimonious exhortations might grate on modern ears, but Bryan was softening the threat that learning science would make women unfeminine.[22]

Following pioneers such as Wakefield and Bryan, more and more women started writing scientific books for women and children. Rather than the abstract, distanced tone adopted in men's texts, these women chose a chatty, personal style and often constructed their books as fictional dialogues, a traditional literary device. As in the imaginary conversation between Euphrosyne and Cleonicus (see Figure 18), an ignorant pupil fed leading questions to a wise teacher. These female writers revolutionised traditional conventions by placing not older men, but older women, in the position of authority.[23]

Priscilla Wakefield made Linnaean botany accessible and permissible for women. Although her style would seem unbearably turgid to modern school children, Wakefield was committed to attracting a teenage audience. Her diary entries show that it was hard work: 'Regular application – made slow progress from the necessity of consulting a variety of books & the difficulty of accommodating my language to my young readers.'[24] She set up a simple if somewhat contrived fictional scenario for her *Introduction to Botany*. When Constance goes off for the summer to visit some cousins, her sister Felicia – left behind, for some unexplained reason

– is rescued from depression by the motherly Mrs Snelgrove, who teaches her botany. No longer lonely, Felicia sends twenty-eight letters to Constance enthusing about the joys of observing and identifying flowers. Accompanying her letters with diagrams and tables, Felicia consoles Constance with a correspondence course in Linnaean taxonomy.

Aware of parental sensibilities, Wakefield cleaned up Linnaean botany by safely sorting plants not into marriages and adulterous relationships, but into nations, tribes and families. Through Felicia, Wakefield carefully described the parts of a plant and systematically went through each class giving examples of familiar English flowers. Relying on Withering's expurgated text, she deftly avoided overt references to plant sexuality and either explained or anglicised technical Latin terms: the female *pistillum*, for instance, became the neutral pointal. Wakefield was selective in her translations. She carefully discussed when petals should be called the corolla, but omitted to mention that for Linnaeus it was a marriage bed. Similarly, she skipped rapidly over *gynandria*, which means the union of woman and man; this class has twenty stamens and one pistil, and so was potentially open to all sorts of sexual interpretations.

This is Wakefield's account of pollination – detailed, precise, yet discreet:

> The pollen or dust, which bursts from the anthers, is absorbed by the pointal, and passing through the style, reaches the germ, and vivifies the seed, which, without this process, would be imperfect and barren. The stamens, pointal, and corolla, having performed their respective offices, decline and wither, making room for the seed-bud, which daily increases, till it attain its perfect state.[25]

This sanitised botany was safe, carried out at home under the protection of an older woman – a mother or a governess. Wakefield emphasised that botany was an ideal topic for girls because it was easy, cheap and encouraged them to take healthy exercise outside rather than idling away their hours gossiping or worrying about the latest fashions. By studying flowers, young women could spend their time profitably, rather than spending their parents' money. By imposing order on nature, they would learn how to conduct their own lives in a more organised manner. 'Order,' Wakefield taught, perhaps with her errant husband and sons in mind, 'should extend to all our concerns; the disposal of our time and money, the proportion of our amusement and business should be regulated by some rule, and not left to the direction of mere chance, as is too often the case with many thoughtless people.'[26]

Wakefield wrote because she needed the money, but her Quaker attitudes remained with her. Philanthropy was a fashionable way of promoting the recipient's virtue as well as the giver's, and Wakefield's charitable savings banks were commercial concerns that imposed moral behaviour on the poorer classes. Similarly, her books earned money but were laced with religious homilies. Felicia smugly relayed her governess's advice to look at plants not just for their botanical interest, but also because they were useful. We should, she told Constance, 'perceive and admire the proofs of Divine Wisdom exhibited in every leaf, and in every flower'.[27]

Although Wakefield approved of hard work, she did not envisage her female pupils undertaking a scientific career. She called one of her other books *Mental Improvement*. This stern title reflected her belief that for girls the point of learning about science was to develop their characters and help them become better wives and mothers. Not for her any recommendations to embark on an unconventional life as a female intellectual. Young girls, Wakefield taught, should not expect to have a deep understanding of science, still less display their knowledge. Under the watchful tutelage of Mrs Snelgrove, Felicia never ventured beyond the garden and the adjacent fields. While happy to regurgitate predigested Linnaean knowledge, she never proposed challenging Linnaeus's system or making new discoveries of her own.

In the early nineteenth century, the market for women's botanical books expanded rapidly. Authors catered for two types of purchaser – mothers reading at home with their daughters, and teachers of the new courses in botany being set up in girls' schools. Wakefield was joined by several other female writers, and some women established successful careers as botanical illustrators. Was this the start of new scientific opportunities for women? In retrospect, encouraging girls to study botany seems, perversely, to have worsened their position.

One problem was that the old associations between winsome women and fragile flowers were being revived. Mary Wollstonecraft objected strongly, cleverly using botanical images to support her case. She regarded this metaphorical bond between women and flowers as a male plot to deprive them of their rights. Men, she sneered in the very first paragraph of her *Vindication of the Rights of Woman*, were cultivating glamorous women like exotic plants – luxury items to be discarded when their beauty decayed. She expressed this power relationship evocatively: 'like the flowers which are planted in too rich a soil, strength and usefulness are sacrificed to beauty; and the flaunting leaves, after having pleased a fastidious eye, fade, disregarded on the stalk, long before the season when they ought to have arrived at maturity.' Instead of reducing themselves to 'sweet flowers that smile in the walk of man', she protested, women should cultivate their brains: 'I do not wish them to have power over men; but over themselves.'[28]

Anything to do with botany became downgraded precisely because it was a field where women flourished (hard to escape those botanical metaphors!). Women were producing accomplished paintings of flowers for technical books as well as for decoration, but their work could easily be dismissed as trivial because it was feminine. As a (male) drawing teacher observed, 'There are men of abilities, who think it beneath them to paint flowers, and affect to treat that branch of the art with contempt.' Some women did become professional flower painters, but this speciality became marginalised as a minor side-line in the world of art. Moreover, expert botanists disdained the elaborate coloured flower arrangements designed for and by women, preferring meticulous drawings in black and white. However skilled their work, and however much in demand, women artists were by definition working in less important genres than their male colleagues.[29]

During the nineteenth century a new divide yawned open between salaried career professionals and dedicated but unpaid amateurs. Men could take advantage of either option, but women had no choice – they were obliged to be amateurs.[30] Collecting, drying and pressing flowers might be suitable activities for leisured ladies, but male natural historians preferred to boost the virility of their image by activities such as hunting for animals or chipping out fossils. In any case, instead of being tied to the home like women, they were free to explore remote regions of the world and bring back new plants. These exotic imports were used to stock Kew Gardens and other commercial testing stations, which were very different from the domestic back garden in which Felicia had wandered under the protection of Mrs Snelgrove.

Chapter 11

Mary Shelley/Victor Frankenstein

The Men, by thinking us incapable of improving our intellects, have entirely thrown us out of all the advantages of education; and thereby contributed as much as possible to make us the senseless creatures they represent us. So that, for want of education, we are render'd subject to all the follies they dislike in us, and are loaded with their ill treatment for faults of their own creating.

Sophie, *Woman Not Inferior to Man*, 1739

It was on a dreary night in November . . . one of English literature's most famous opening lines. It first appeared in 1818 to introduce the dramatic scene when Victor Frankenstein infuses life into the creature he has assembled from illicitly gathered body parts.[1] They were the first words Mary Shelley (1797–1851) committed to paper after a troubled night of waking nightmares that inspired the plot for her extraordinary novel, *Frankenstein*. Confined to a house in Switzerland by the rainy weather, Shelley and her travelling companions were involved in a competition to terrify each other by telling ghost stories. To her embarrassment, she could think of nothing suitable to hold the group's attention. Late into the night their conversation had ranged over recent experiments and discoveries – the new electric batteries that galvanised corpses into action, Erasmus Darwin's desiccated creatures that wriggled vigorously after soaking. Was it possible, the friends wondered, that science could solve the biggest question of them all, the nature of life itself? When Shelley eventually went to bed, her imagination feverishly conjured up a horrific scene. At last she had a tale that would enthral her friends.

Although created less than 200 years ago, Frankenstein has become a mythical figure. Everyone knows his story – or do they? Mary Shelley's original book, first published in 1818, was very different from modern versions. Shelley herself rewrote substantial sections, cutting out a lot of the science and adding sentimental chunks to soften the punch of her message. A century later, Boris Karloff created the definitive film icon of a hatchet-faced monster with bolts through its neck – but this was a parody of Shelley's sensitive progeny racked by remorse. More recently, the publicity for Kenneth Branagh's production advertised its authenticity, even though there were substantial changes in the plot. This is a brief outline of the first edition of *Frankenstein* . . .

> Like a set of Russian dolls, in *Frankenstein* three people's stories are embedded one within the other. The outermost wrapping concerns Robert Walton, a polar explorer who rescues an emaciated, near-frozen stranger and brings him back to life. Unlike later episodes, this is a normal process of revival, using brandy, warmth and food. The narrative of this unexpected visitor, Victor Frankenstein, forms the next layer in . . .
>
>> Here the resuscitation is more artificial and sinister. After studying different types of science, Frankenstein plunders dissecting rooms and charnel houses to construct an eight-foot human replica, whom he mysteriously transforms into a living being. This had been Shelley's night-time inspiration – the creator kneeling appalled as his horrendous handiwork stirs into life. Rushing from the room in disgust, Frankenstein abandons science, succumbs to weeks of fever, and then struggles to resume normal activity. But six months later, the nightmare erupts again. The creature kills Frankenstein's little brother, and a servant is unjustly hanged. Frankenstein is horrified, but when he discovers his daemon, consents to hear his version of what has happened, which comprises the core of the book . . .
>>
>>> Misery and rejection, not innate evil, drove the fiend's monstrous behaviour. Sheltering next to a cottage, he had secretly helped a humble family to survive and, by eavesdropping on their conversation, learned first how to talk, and then how to read. His ghastly appearance conceals a truly civilised character, yet he repulses

everyone he tries to befriend and despairs of being condemned to a life of isolation. Murder is his weapon of revenge, a way of blackmailing Frankenstein into creating a female companion to lighten his lonely existence. He extorts the promise of a partner from his listener, who resumes his own story . . .

Although Frankenstein does start to collect the ingredients for a woman, he destroys his half-finished work in disgust. The avenging creature kills first his closest friend, and then – deprived of a wife for himself – strangles Frankenstein's beloved cousin Elizabeth on the night of their marriage. Racked with self-hatred, Frankenstein dedicates himself to killing the fiend that he has generated. Ineluctably bound to one another, the monstrous couple roam across Russia and up into the Arctic north, until they are separated by a crack in the ice. Hovering at the edge of death, Frankenstein now persuades his own listener, Walton, to pursue the creature and kill him . . .

Under pressure from his crew to turn southwards to safety, Walton breaks his promise to Frankenstein, who dies. The book closes as Frankenstein's demonic progeny, now a martyr who repents of his evil deeds, disappears northwards across the icy waves to sacrifice himself on his own funeral pyre.

Frankenstein is fiction, but it is also a historical document packed with information about attitudes towards science and women in the early nineteenth century. As one of Shelley's first reviewers commented, *Frankenstein* 'has an air of reality attached to it, by being connected with the favourite passions and projects of the times'.[2] Together with its countless reinterpretations, *Frankenstein* is one of the most heavily studied novels in the English language, and yet – like all enduring myths – there is always something new to learn from it.

Where do the boundaries lie between science and imagination, between fact and myth? The distinctions are blurred because there are different sorts of truth. Mythical heroes may engage in imaginary exploits, but their moral dilemmas are rooted in real-life ethical choices. Conversely, factual accounts of the physical universe sometimes turn out to be less objective than scientists claim. Many of the events in Shelley's book could not possibly have happened, but, like fairy stories, they seem to take place in an adjacent world that resonates with our own hopes and anxieties. Her larger-than-life actors are slightly unreal, yet their unfolding story carries warnings about how our own lives should be conducted.

Shelley could not have foreseen how her novel would be twisted to fit scary scenarios lying ahead in the future. However, she certainly did appreciate the power of mythical stories. Classical legends were far more familiar then than now, and the basic characters and plots were instantly recognisable. The full title Shelley chose for her scientific fiction was *Frankenstein; or, the Modern Prometheus*. Her contemporaries knew that Prometheus had made the figure of a man from clay and then, with the help of Athene (Minerva), had stolen the secret of fire from the gods so that he could infuse his statue with life. They also remembered that Prometheus was the grandson of Uranus, the planet discovered by William Herschel – an especially important link for Shelley, who credited her beautiful complexion and fine haze of hair to her birth beneath one of Caroline Herschel's comets. Romantic poets often identified themselves with Prometheus as a saviour/creator figure. Particularly during the French Revolution, Prometheus became a symbol of rebellion against tyranny; it was only later that he came to represent a warning against human ambition.

Conservatives condemned *Frankenstein* because it frightened them. In the Bible, it was God, not Frankenstein or Prometheus, who had made Adam out of dust and breathed the fire of life into his nostrils. Scientific research promised so many rewards, yet surely it was sacrilegious to suggest that a mere mortal could create life? Shelley's *Frankenstein* still evokes people's deepest fears of disease, revolution and disaster, and it also seems to foretell future scientific catastrophes. The story remains so powerful because, like the myths on which it is based, it has repeatedly been retold. Resembling other fictional characters who verge on reality – Faustus, Don Juan, Sherlock Holmes – Victor Frankenstein and his creature have taken on lives of their own as, time and again, they have been reinterpreted. Stripped of Shelley's original subtleties, the evil scientist and the demonic monster have symbolised one threatening evil after another – Irish nationalism, the atomic-bomb project, transplant surgery, and now the 'Frankenfoods' produced by genetic modification.[3]

When Shelley created Frankenstein, the modern Prometheus, she wrote a scientific morality tale, the modern equivalent of a religious parable.

She may also have been thinking about the second part of this story, familiar then, but now less well known. Zeus wanted to punish the human race for Prometheus's transgression, and he himself gave life to a clay statue. Zeus's creation was not a man, but a woman – Pandora. At her birth, the Olympian gods taught Pandora all the female virtues, but when she was presented as a gift to Prometheus's brother, Pandora lifted the lid from her giant vase and unleashed evil throughout the Earth. Only Hope remained inside her casket.

Shelley was strongly influenced by her father, the radical reformer William Godwin. When she was little, he used her and the other children of the household as test readers for his books. In one of these, he outlined the linked Prometheus/Pandora story. 'The fable of Prometheus's man, and Pandora the first woman, was intended to convey,' he wrote, 'to how many evils the human race is exposed; how many years of misery many of them endure . . . how many vices are contracted by man, in consequence of which they afflict each other with a thousand additional evils, perfidy, tyranny, cruel tortures, murder, and war.' Modern readers forget how closely the stories of Prometheus and Pandora are tied together, yet Godwin's summary reads like an outline of his daughter's Frankensteinian nightmare.[4]

Victor Frankenstein was a figment of Mary Shelley's imagination but, like real-life women/men couples, their partnership reveals contemporary attitudes towards women and science. Shelley was well informed about the latest scientific debates, and *Frankenstein* was an astute study of the society in which she lived. By the end of the eighteenth century, science had become a fashionable topic, and educated readers – women as well as men – were knowledgeable about at least the basic principles.

Even though female authors were unable to embark on active careers within science, their educational books were enormously important because they presented scientific knowledge in a way that everybody could understand. As a result, science was no longer an arid esoteric subject limited to a privileged few, but was an important part of nineteenth-century culture. During the eighteenth century, men had cashed in on female wealth by writing patronising, simplified books that treated women as consumers of science. By the middle of the nineteenth century, some women were supporting themselves (and their dependants) by writing scientific books read by both the sexes. Education seemed to be suitable work for women, who were not deemed sufficiently original to generate their own ideas, but were at least clever enough to understand and explain the achievements of others.

Conventional scientific texts and imaginative works of literature are often seen as lying on either side of an impermeable boundary. More realistically, they can be viewed as marking opposite ends of a continuous spectrum. Ethical problems of science were aired not only in political debates and learned journals, but also in fiction. Reciprocally, scientific authors devised fictional settings to engage their pupils' interest. Priscilla Wakefield's *Introduction to Botany* certainly wasn't a novel, but neither was it a dry collection of facts. By inventing characters such as Felicia, Constance and the motherly Mrs Snelgrove, she was – like other writers – blurring the division between fact and fiction. Spread out between the fact/fiction poles lay educational books like Wakefield's and Bryan's that relied on fictional scenarios, and also fictional books that relied on scientific arguments – novels like Maria Edgeworth's *Belinda* (which included another dig at Thomas Day's educational experiments) and *Frankenstein*.[5]

Although there were plenty of opportunities for women to learn about science, they were not encouraged to embark on laboratory experiments. As a woman, Shelley was restricted to observing science rather than participating in it, but she did learn science from her husband Percy. He had been a keen experimenter while he was a student at Oxford and still kept up with the latest research. For his twenty-sixth birthday, she gave him a hand-stitched balloon and a telescope, which they bought together in Geneva. A voracious reader, she borrowed his textbooks and kept up-to-date with scientific issues being debated in the journals. However, whereas Percy Shelley had had hands-on training at school, this experience was denied to Mary Shelley.

Shelley's father planned to teach her 'some smattering of geography, history and the other sciences', and for part of her childhood Shelley studied these more technical subjects at home with a governess and a tutor. Specific details have not survived, but her teachers would almost certainly have given her some of the little conversational books on science being written explicitly for girls by Wakefield, Bryan and other female authors.

And for chemistry, the most likely fictionalised account for Shelley to have read was Jane Marcet's *Conversations on Chemistry*.[6]

Jane Marcet, the daughter of a wealthy Swiss banker, was one of the most influential scientific writers amongst the generation after Wakefield. Instead of botany, the conventional safe science for women, she chose to write about chemistry, one of the sciences central to *Frankenstein*. Marcet set up her books as cosy dialogues between Mrs B and two young girls. (Although the link is probably coincidental, Margaret Bryan's frontispiece (Figure 28) does seem marvellously apt.) By fictionalising her presentation, Marcet created new audiences for science and so helped to establish its importance.

When Mary Shelley was writing *Frankenstein*, for chemical information she chose the same scholarly source as Jane Marcet – Humphry Davy, a young lecturer at the newly founded Royal Institution (Figure 29). In James Gillray's caricature, Davy menacingly clutches a pair of bellows and hovers behind the professor who is forcing his exploding victim to inhale laughing

Fig. 29
Scientific Researches! – New Discoveries in Pneumaticks! – or – an Experimental Lecture on the Powers of Air.
Hand-coloured etching by James Gillray, 1802.

gas (nitrous oxide) from a series of flasks. Davy attracted hundreds of people to his lectures, but the savagery of Gillray's humour indicates how tenuously Davy clung to his position. Traditionally, chemistry was the poor relation of physics and mathematics, disparaged as a practical rather than a theoretical subject. In addition, since Lavoisier's revolutionary changes, chemistry had become associated with political radicalism and danger-ously progressive thinking.[7]

Davy wanted to demonstrate how new chemical techniques could probe deep into the nature of matter, but his critics accused him of introducing too much spectacle. Unlike the Royal Society, the Royal Institution was – in principle, at least – open to the public, and Gillray has satirised the members of the audience as cruelly as the performers. The lower-class people to the left gawp in amazement, while the fashionably dressed couple at the front are assiduously taking notes. Conservative observers sneered at these educational aspirations. Samuel Taylor Coleridge worried about the slippery slope that would result in debasing knowledge by making it available to all. 'You begin, therefore, with the attempt to *popularise* science,' he warned – 'but you will only effect its *plebification*.' Coleridge's friend Robert Southey divided the spectators at the Royal Institution into two groups – bored men and superficial women. 'Part of the men were taking snuff to keep their eyes open,' he reported, 'others more honestly asleep, while ladies were all upon the watch, and some score of them had their tablets and pencils, busily noting down what they heard, as topics for the next conversation party.'[8]

Although Marcet might seem to be unimportant because she was a chil-dren's writer, her work had vital impacts on nineteenth-century science. She was not a passive transmitter of diluted science. On the contrary – by focusing on chemistry for women, she changed the ways in which science itself was understood and performed. In her chatty little books, a woman and two girls practise chemistry at home, the conventional place for women. By creating this fictional setting, Marcet converted Davy's showy demonstrations into a safe domestic activity that anyone could tackle. Davy made chemistry appear exciting, dangerous and politically risky – definitely a subject for men rather than women, and far from being a sober science. But by making chemistry respectable, and even safe enough for girls, Marcet helped to make it central. No longer relegated to the scientific margins, chemistry became a subject that enabled far more people than ever before to participate in scientific investigations.

Married to a Fellow of the Royal Society, Marcet belonged to a group of educated women who transformed science communication. Many of them the wives and daughters of eminent scientific men, they were related to one other in extended intellectual networks resembling royal dynasties. These women were vital for the development of science in the early nineteenth century. Marcet was very friendly with Maria Edgeworth, who had a special reason to be grateful – when her sister accidentally swallowed some poison, Edgeworth remembered Marcet's lessons in *Conversations on Chemistry* quickly enough to prescribe a life-saving chemical remedy. Marcet was also close to Mary Somerville, whose lucid explanations of complex French physics made the latest research accessible to science students – men as well as women. Rather like the literary bluestocking circles of the eighteenth century, women like Mary Somerville, Jane Marcet, Mary Lyell and Maria Edgworth formed their own close female groups, exchanging scientific news and providing each other with practical and emotional solidarity.[9]

Marcet's *Conversations on Chemistry* was a huge success. The book first appeared in 1806, but ran through sixteen editions during the next forty years and was twice translated into French: in the United States alone, 160,000 copies were sold. Many thousands of scientists must first have learned their chemistry from Marcet, but her most famous pupil was Michael Faraday, founder of electromagnetism and Davy's successor at the Royal Institution. As a young bookbinder's apprentice, Faraday taught himself chemistry by reading Marcet's *Conversations* in the evenings after work. Long after he became one of England's most prestigious scientists, Faraday sent Marcet copies of his scholarly articles and praised his first teacher for giving him great pleasure at the same time as writing accurately. By capturing his imagination, Marcet had drawn Faraday into the world of facts. Marcet initially wrote for women, but – like other female scientific authors – she dramatically affected the very nature of science itself.

Nevertheless, being a scientific woman demanded a thick skin. By presuming to enter the masculine territory of science, women were transgressing conventional boundaries. It was not only men who drew intellectual distinctions between the sexes – many women also worried that engaging in academic activities would detract from their femininity. 'All women possess not the Amazonian spirit of a Wolstonecraft,' campaigned one of her female critics, Mary Radcliffe. She warned that overturning male oppression might entail 'throwing off the gentle garb of a female, and assuming some more masculine appearance'.[10] Correspondingly, pictures portrayed Minerva, the goddess of wisdom, as a muscular, heavy-set woman – a masculine goddess to indicate that learning was associated with men.

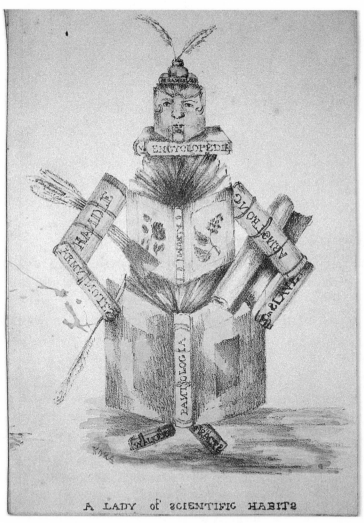

Fig. 30
A Lady of Scientific Habits.
Lithograph, early nineteenth century.

New images appeared to back up the moralising books, articles and lectures that prescribed appropriate behaviour for women. In the frontispiece to her book on astronomy, Bryan had shown herself holding a pen (Figure 28), but the print reproduced in Figure 30 mocks female pretensions to authorship. This writer's inkwell is ludicrously planted on top of her head, and a sheaf of quills is precariously balanced beneath her arm. 'The lady of scientific habits' is a travesty of femininity. Drawn in straight lines rather than curves, she wears a skirt compiled from *Pantalogia*, a twelve-volume encyclopaedia and here a pun on men's trousers. The other tomes also play verbally on books and the body – Walker's *Tracts* for her feet, Armstrong's *On Slavery* for her left arm, *Army Notes* for the other. In place of her breasts are two flowers, which indicate the contents of the Album covering her chest. Far from being an advertisement for scientific botany, this lady was herself designed to be stuck inside an album, along with pressed flowers, drawings and verses. This whimsical parody of intellectual dedication signalled to women the unfortunate consequences of devoting themselves to scholarship.[11]

The old stereotypes were still being reinforced: scientific learning for men, domestic skills for women. Although women were campaigning for better female education, their demands were very different from those of modern feminists. Most of them agreed that women's minds were different from those of men, inherently less suited to the methodical work required in science. 'We are not formed for those deep investigations that tend to the bringing into light reluctant truth,' maintained the novelist Laetitia Hawkins. She backed up her argument by pointing out the unflattering effects of concentrated thought – a furrowed brow and fixed gaze lent dignity to a man, but could only detract from a woman's soft features. Marriage and motherhood were sacrosanct even for Wollstonecraft, while educators like Wakefield and Bryan stressed that a woman's place lay at home. 'Each sex has its proper excellencies,' observed the conservative Hannah More, as she fought to retain the *status quo*. Surely, she asked, women should continue in their allotted roles, rather than challenge God's will by attempting to emulate the work of a man?[12]

Hardly surprising, then, that Marcet and other women decided to publish their scientific textbooks anonymously. At first, *Frankenstein* also appeared anonymously, although it did bear two men's names – William

Godwin and Percy Shelley. Mary Shelley dedicated her book to Godwin, her father, which might have given a strong hint of the author's identity. But the preface was written by Percy Shelley, and many readers inferred that he had written the whole book – including Sir Walter Scott, to whom she proudly revealed the truth.

Some of the early reviews were vicious. Shelley left England soon after the publication date, so hopefully she never read that her book was 'the foulest Toadstool that has yet sprung up from the reeking dunghill of the present times'. But sales were high, and five years later the second edition did carry her name on the title-page. Critics were astonished at this female feat. 'For a man it was excellent,' exclaimed one reviewer – 'but for a woman it was wonderful.'[13]

Literary analysts often hail Mary Shelley's book as the first in a new kind of literature – science fiction. *Frankenstein* was not, of course, intended to be a documentary. Nevertheless, it was solidly based on cutting-edge scientific research into polar exploration, electricity and chemistry. Shelley's novel was shocking precisely because it wavered on the edge of feasibility. Modern alarmists celebrate *Frankenstein* for its prescient warnings of subsequent scientific disasters and dilemmas, but Shelley produced a commentary on the present as much as a manifesto for the future.

Using fiction, Shelley explored the doubts hovering around science's respectability. Scientific men felt insecure, unstable: they had no fixed identity, no career structure, no professional organisations to support and advise them. Sceptics questioned whether ambitious scientific projects could be reconciled with the orderly behaviour and clear, logical thought expected of a gentleman. Conventional scholarship belonged in gentlemen's studies, and was (supposedly) pursued for the sake of abstract knowledge and human welfare. Scientific research, on the other hand, entailed travelling overseas and demanded manual work of the type formerly reserved for labourers and servants. Some men even expected to be paid for their work or rewarded for their inventions! Educated women presented another challenge. Surely allowing them to engage in science would threaten its exclusivity, would diminish its appeal as a prestigious activity for privileged men?[14]

The three layers of *Frankenstein* probed the public value of scientific research from different angles. Reading the novel's outer framework, it is clear that Shelley had been keeping up with the latest scientific discussions about magnetism and polar voyages. Foreign travel boomed after Napoleon's defeat in 1815, and the press was packed with articles discussing the advantages of Arctic exploration. Her fictional Walton personified a new type of celebrity – the scientific explorer. In the past, military champions had been hailed as the nation's heroes. Now their place was being taken by adventurous young men who set out to conquer the world by gaining knowledge rather than taking over territory and people.

Walton romanticised the aspirations of scientific investigators and also of naval administrators. Far from being the disinterested pursuit of pure knowledge, scientific exploration was indelibly tied to commercial and imperial profit. Propagandists advertised the importance of scientific research for improving navigation and exploiting the resources of overseas colonies. Government, commerce and science went hand in hand as Britain expanded her empire. As a leading Admiralty official put it, searching for the North Pole was essential for extending 'the sphere of human knowledge'; and, he continued, 'Knowledge is power.'[15] *Knowledge is power*: a direct quote from Francis Bacon, and one that made the goals of modern science sound dangerously like the dreams of those ancient mystics and alchemists whose works Frankenstein had studied with such fascination.

Scientific research was being promoted for its financial rewards, but contemporary questions about the moral cost of progress thread through the central sections of *Frankenstein*. At the newly founded Royal Institution, Davy – in an expression reminiscent of Bacon – had boasted that experiments enabled a man 'to interrogate nature with power, not simply as a scholar, passive and seeking only to understand her operations, but rather as a master, active with his own instruments'. For him, the electric battery was not only a miraculous new source of energy, but also 'a key which promises to lay open some of the most mysterious recesses of nature'.[16] Like Bacon's earlier followers, Davy wanted not only to learn about female nature, but also to search out her secrets and control her.

But Davy also cautioned against scientific speculators who promised too much. 'Instead of slowly endeavouring to lift up the veil concealing the wonderful phenomena of living nature,' he warned, 'they have vainly attempted to tear it asunder.'[17] *The veil of nature*: an old and familiar image.

How could scientific experimenters gain prestige by promising rewards from their research, yet avoid ripping away the veil like over-confident magicians? When she evoked Frankenstein's inner turmoil, Shelley tapped into these subterranean anxieties.

Davy was important to Shelley's writing both for his scientific ideas and for his iconic status. She studied his work while she was creating *Frankenstein*, and she must also have come across many references to him in other books and articles; she may perhaps have attended one of his lectures at the Royal Institution. Davy was England's leading expert on electrochemistry. An energetic self-promoter, he advertised how electric batteries, which had only recently been invented, could reveal new elements and pick apart the chemical processes that were fundamental to life.

Unlike older scientific topics, such as mechanics and mathematics, electrochemistry was new and exciting, tinged with political threats as well as experimental risks. In one of his lectures that Shelley studied, Davy sketched out a Frankensteinian vision. Electrical research, he claimed, has not only shown the chemist how to understand the world, but has also 'bestowed upon him powers which may be almost called creative; which have enabled him to modify and change the beings surrounding him'. Other chemists also wrote with messianic fervour. 'We are now admitted,' enthused an eminent Scottish professor, 'into the laboratory of nature herself.'[18]

Experimenting in his own laboratory of nature, Frankenstein discovered the secret of life. As with polar exploration, Shelley was not just fantasising about science, but was fully up-to-date on contemporary issues. While she was writing *Frankenstein*, the old debates about the nature of life had reached a particularly acrimonious peak. At one extreme lay the materialists, who insisted that life lay in matter itself and was just a question of somehow rearranging fundamental building-blocks. Their opponents objected to any such suggestion that life could originate in a chemical process. Surely, they argued, life cannot just be a result of how atoms and cells are organised, but must depend on some soul or spirit that is infused by God. John Abernethy, a prominent London surgeon, had recently proposed a compromise between these two opposing camps in the vitalism controversy, the scientific reductionists and the religious traditionalists. Perhaps, he suggested, the source of life might be some superfine invisible substance, analogous to electricity.

Electricity – could this be the secret of life itself? It was a captivating idea. Since Shelley was a small child, doctors had been claiming that dead people could be revived by passing strong electric currents through them, and a few successes were reported after cases of drowning. At last it seemed as if the medical dream of resuscitating the dead might be attainable. In one particularly famous London experiment on a newly hanged criminal, spectators watched as the corpse's eye leered open, his clenched fist rose into the air and his legs kicked violently while his back arched. Similarly, Shelley described how, using his 'instruments of life', Frankenstein managed to 'infuse a spark of being' into the inert assemblage of limbs and organs that he had sewn together; he watched 'the dull yellow eye of the creature open; it breathed hard, and a convulsive motion agitated its limbs'.[19]

One of Abernethy's most outspoken critics was his former pupil William Lawrence, who was a friend of the Shelley couple as well as their medical adviser. Accusing Abernethy of fudging the issues, Lawrence ridiculed his suggestions that electricity could stand in for the human soul. Whether superfine or coarse, insisted Lawrence, matter is matter, and that is where life is to be sought – there is no magic moment of creation when a mysterious vital force is somehow infused into inert material. Shelley signalled her allegiance to Lawrence by portraying Frankenstein – representative of the Abernethy faction – as a blundering experimenter whose ambitions failed.

In telling her story about Frankenstein, Shelley ruminated on the fate of science and peopled her fiction with scientific stereotypes. Shelley's only survivor is Walton, the romantic dreamer who has renounced poetry and turned instead to scientific exploration. With soldierly dedication, he has toughened his body and trained his mind before setting off with child-like curiosity to investigate the frozen north. Although fired with utopian ecstasy, unlike Frankenstein, Walton does not tamper with the powers of nature but contents himself with uncovering them. As he searches for hidden waterways and the source of the Earth's magnetism, Walton is driven by the same noble aim of glory that initially inspired Frankenstein. Glory, not wealth, is his objective, and he is willing to sacrifice himself for the benefit of all humanity.

Studying at home as a child, Frankenstein had been inspired by reading sixteenth-century mystical chemists; like Walton, he was fired by starry-eyed visions of developing a universal panacea. At university, Frankenstein encountered two very different models of scientific investigator. The short and ugly Krempe, a parody of Enlightenment systematisers, ridiculed ancient knowledge and prescribed a rigorous modern reading list of texts in natural philosophy. Rather than striving for power and immortality, the cramped Krempe insisted on pinning nature down in the minutest detail and squeezing her into the confines of mathematical laws.

In contrast, Waldman was a far more sympathetic teacher. A charismatic chemist with a sweet voice and a sprinkling of distinguished grey hairs, he presented a progressive view of science. Frankenstein listened enthralled as he explained how modern chemists had inherited the alchemists' dreams. Echoing Bacon and Davy, Waldman portrayed chemists as the miracle-makers of a new scientific age, who 'penetrate into the recesses of nature, and shew how she works in her hiding places . . . They have acquired new and almost unlimited powers.' Nevertheless, the dignified Waldman was governed by his intellect rather than his imagination. If you want to become a real man of science, he advised Frankenstein, then you must study not just chemistry, but all the scientific subjects, including mathematics.[20]

Unlike these two mundane teachers, Frankenstein is never physically described, as though he were pure mind, a restless enquiring soul existing on a different level. Born into a respectable orderly family, he is aware from childhood that he is destined for great things, and he ranks himself above 'the herd of common projectors'. Like a Romantic poet, he is 'a celestial spirit' whose soul is elevated by contemplating the beauties of nature – an echo not just of Shelley's husband Percy, but also of Davy, who wrote poetry as well as performing chemistry.[21]

So long as science and imagination are kept apart, Shelley's fictional world remains an orderly place. Dangers arise only when the boundary melts, and imagination diffuses across into the scientific realm. Instead of behaving like a sober natural philosopher, Frankenstein lets himself be flipped over into dreams and fantasies. No longer a respectable man of science, he becomes a mad genius, governed by his internal creative fire as though he were a poet or an artist. Davy was London's leading chemist, yet he also regarded himself as a genius – a creative investigator rather than a plodding natural philosopher. Science, he taught, was far more than a series of mechanical operations because it involved the higher planes of the human soul. By using scientific techniques and instruments, an inspired genius could be at one with the powers of nature and could uncover the secrets of life, thought and morality.

Walton the heroic explorer and Frankenstein the mad genius. Shelley contemplated the fluid character of contemporary science and crystallised out two heroic role models. Walton survives but fails, defeated by the strength of a female nature who refuses to let him approach the North Pole, source of the electrical and magnetic energy that activates the world. Frankenstein, far more presumptuous than Walton, aims not only to uncover nature's secrets but also to control her powers – an over-ambitious project that is bound to fail.

In common with Shelley, modern female authors choose science fiction to express not only their reservations about science, but also their dissatisfaction with society. At the core of her book, Shelley used Frankenstein's creature to expose the shortcomings of modern civilisation. 'Some have a passion for reforming the world,' she told her diary (she meant her parents and her husband). 'I respect such,' she went on, 'but I am not for violent extremes, which duly bring on an injurious reaction.' Rather than producing fierce political tracts like other members of her family, Shelley preferred the more enigmatic genre of fiction.[22]

Again, she drew on current scientific debates. Shelley's journals show how avidly she read, novels as well as scholarly works on politics, philosophy, history. Perhaps she sometimes felt like the 'Lady of Scientific Habits' (Figure 30) – composed entirely of books and devoid of a body, this desexualised woman can give birth to nothing but more books. Was *Frankenstein* a substitute offspring, composed as it was while she was pregnant and haunted by the death of her first baby? In her book, Shelley explored the bond between the misshapen progeny and its arrogant creator. Viewing the world with a child-like naivety, the abandoned and nameless creature soon experiences the violence meted out to ugly outsiders. Like a magnifying mirror, his distorted image reflects the hierarchies of power that kept men at the centre and forced others – non-Europeans, women, labourers – to the margins.[23]

Like many of her readers, Shelley knew about Erasmus Darwin's theory of evolution, his optimistic idea that species gradually improve by adapting to their surroundings over long periods of time. She was also familiar with Lawrence's controversial suggestion that superior human beings could be produced – like agricultural animals – by selective breeding. During the eighteenth century, several so-called wild boys and girls had been discovered roaming the countryside, and the problems of educating and assimilating these children fuelled intense discussions in books and newspapers.

Shelley was familiar with the questions these children raised. Were they truly human, or had they deteriorated beyond recall? Were there fundamental differences between people and animals? And how should non-Europeans be classified – were they hopeless savages, or could they be brought up to Western levels? Linnaeus had even identified a distinct human species, *homo ferus,* a mute and hairy beast that shuffled on all fours. Like *homo ferus,* Shelley's creature is prodigiously strong and can withstand extremes of temperature. Eavesdropping in the woodland cottage, he goes through a crash course in Western culture, and yet – like a feral child – he is never accepted by human society.[24]

Shelley was also interested in another type of classification – the differences between men and women. However trenchant her critique of men's scientific aspirations, in many ways Shelley reiterated traditional attitudes towards women, nature and science. Even though her own life was far from conventional, she portrayed women as intrinsically fragile, home-loving creatures, governed by their emotions and their imagination rather than by rational thought. In contrast, her fictional men were active, outgoing, committed to academic study and capable of long feats of endurance. Shelley's women conform to expectations by excelling in social skills rather than the rigorous intellectual work needed for scientific research – Elizabeth is the only one who can communicate with the servants.

As well as the myth of Prometheus, perhaps she was also thinking about Pandora's birth, when the deities flocked round to teach her 'every female Art', and the muscular Minerva dwarfed the delicate Pandora (see Figure 3). Rather than imparting scholarly wisdom to Pandora, Minerva (Athene) consolidated female stereotypes by handing her protégée a shuttle to symbolise 'the secrets of the Loom'.[25] In *Frankenstein,* the two cousins Victor and Elizabeth are brought up together, but even as children they display characters dictated by convention. They are complementary partners –

Victor is calm, dedicated and obstinate, the ideal counterbalance for Elizabeth's docility, gayness and compassion. He is clearly destined for scientific research, whereas Elizabeth (like Shelley herself) thrives on art and literature. As Victor Frankenstein explains, 'I delighted in investigating the facts relative to the actual world; she busied herself in following the aërial creations of the poets. The world was to me a secret, which I desired to discover; to her it was a vacancy, which she sought to people with imaginations of her own.'[26]

Female nature plays an important role in *Frankenstein*. Following tradition, Shelley personified nature as a woman with a dual personality (see Figure 11). Storms and flashes of lightning accompany episodes of evil, while the placid Swiss countryside restores Frankenstein to sanity. As children, Victor was calmer than his vivacious cousin, but as adults, Elizabeth's serenity contrasts starkly with his feverish delusions and ravaged deterioration.

Condemned to live in domestic environments, Shelley's women are doomed by their emotional self-sacrifice and devotion to duty. Frankenstein's mother dies because, ignoring the cool voice of reason, she insists on nursing Elizabeth through scarlet fever. A serving girl confesses under duress and, with only a female advocate to plead her case, is executed for a crime she never committed. After years of faithful self-denial, Elizabeth is murdered because she obediently enters the bridal chamber alone. The major male characters, on the other hand, leave their homes to pursue active careers. Writing from the polar regions, Walton warns his home-bound sister that he may disappear for years – or even for ever. The close friends Frankenstein and Henry Clerval, in some ways Janus aspects of Percy Shelley, both abandon their families and devote themselves instead to study at a foreign university.[27]

Clerval, Frankenstein explains, 'was no natural philosopher. His imagination was too vivid for the minutiæ of science.' Even though obliged to opt for languages, as a man Clerval goes to university where he can work with scientific detachment and discipline. He chooses classical and oriental languages, which demand a good grasp of complex grammar and can be mastered by studying books. Like English, these are manly languages. Even the creature, eavesdropping on lessons inside the forest cottage, learns English more quickly than Safie, the family's female pupil. Women were taught through conversation rather than books of grammar, and were being steered towards French, judged easy and slightly frivolous.[28]

Pandora and Eve were the traditional sources of sin. However, Shelley broke with convention by making a man rather than a woman responsible for the consequences of human presumption. The campaigner Mary Wollstonecraft, who was Shelley's mother, had written, 'Man has been held out as independent of His power who made him, or as a lawless planet darting from its orbit to steal the celestial fire of reason; and the vengeance of Heaven lurking in the subtile flame, like Pandora's pent-up mischiefs, sufficiently punished his temerity, by introducing evil into the world.'[29] Initially Frankenstein succeeds in usurping the female gift of procreation, but because of his audacity a new Reign of Terror is unleashed across the Earth – he has unclasped a Pandora's box of evil consequences.

On the other hand, Shelley did not take the next step in role reversal. Although she seems to be condemning researchers who try to govern nature by ripping away her veil, in Shelley's book science remains a pursuit for men. Nowhere does she suggest that science might fare better in the hands of women. For the first time, women were being given a scientific education, but they were still excluded from active research. As science started to move out of private homes, men like Davy mounted the public stage as science's star performers. In contrast, his female readers, women like Marcet and Shelley, were restricted to vital but less visible tasks.

Like the women in her own novel, Shelley observed and commented, but did not participate in the manly world of science. A hundred years later, women would be working inside laboratories, studying at universities and demanding to be treated equally, but Shelley seems not to have contemplated this possibility. Like many radical writers, she remained snared in the conventions of her time. Mary Shelley conjured up visionary scenes of scientific horror, yet apparently she could not envisage such a drastic transformation of her own society.

Epilogue

So finally, to end, a scientific joke. The scene is America in the early 1950s. Scientists from the University of California at Berkeley proudly announce that they have created two new radioactive elements – berkelium and californium, atomic numbers 97 and 98. Why, ask the pundits, did they not label them universitium and offium? Then they could have saved the names californium and berkelium for 99 and 100, the next two elements they are destined to make as science advances.

Even though all the natural elements had been isolated, scientists were continuing to create more and more artificial elements. Like collectors of unusual objects, they could not contemplate the possibility that their collection was complete, that their life's work – their obsession – was over. They had drawn away the veil and uncovered nature's secrets, but they did not want the story to end. If they went on striding forwards, then perhaps their triumphant tale of scientific progress could roll on uninterrupted into the future.[1]

Whether it's radioactive atoms or rare animals, to collect is to own, to dominate, to tell a particular story about what's important. Building up a collection gives a sense of power and control because its owner has reorganised the environment. Like scientific detective work, collecting involves searching and classifying. In both, similar problems need to be resolved, similar questions demand to be answered. What is the best way of ordering my discoveries? How can I hunt down an elusive object? Should I keep my colleagues informed, or should I guard my possessions, my knowledge, against my competitors?

An imposing Minerva (Figure 31) dominates the frontispiece of a large

Fig. 31
Minerva displays her cabinet collection of women.
D. Allan, engraved frontispiece of R. E. Raspe, *A Descriptive Catalogue of a General Collection of Ancient and Modern Engraved Gems, Cameos as well as Intaglios, Taken from the Most Celebrated Cabinets in Europe; and Cast in Coloured Pastes, White Enamel, and Sulphur, by James Tassie, Modeller* (1791).

eighteenth-century advertising catalogue for expensive cameos and imitation gems designed explicitly to cater for the tastes of wealthy collectors. Her study is cluttered with desirable objects, many of them strange females. The snake-wreathed head of Medusa stares out from her shield, propped up on a decapitated head. Nearby stands the goddess of medicine, identified by the caduceus on the table leg. Even Minerva herself is a rarity, an unusual woman with muscular arms, her breasts squashed flat by her military armour. Holding open the door to her cabinet of curiosities, she entices customers to imagine the delights concealed within. Minerva's visitors can glimpse some tantalising figures, most prominently a statuette of yet another goddess, the multi-breasted Diana of Ephesus. During the Enlightenment this ancient symbol of fertile nature was a familiar motif, copied from Roman coins and shown on fountains and public monuments. She represented the nurturing mother, a far more common female stereotype than the learned woman.[2]

This image of Minerva would also be appropriate for written collections of women. No accident that in 1706 Johann Eberti chose the title *Open Cabinet of Learned Women* for his book of several hundred female biographies. Just as Minerva tempts purchasers with the fine figurines in her cabinet, so Eberti encouraged readers to marvel at his female intellectual curiosities. Over the centuries, biographers have gathered together famous women who have little in common except their sex. Resembling delicate pieces of porcelain protected within a display cabinet, they are arranged alongside one other within the covers of a book. Their major entry qualification is their womanhood rather than their achievements.

Different collectors have different selection strategies. The first set of biographies devoted exclusively to women was by the fourteenth-century Florentine writer Giovanni Boccaccio. Although his *Decameron* is now far better known, Boccaccio's *Famous Women* was a great success, circulating in hand-copied manuscripts as well as printed editions. He collected more than a hundred subjects, many of them mythological heroines such as Minerva. He chose them not for their abilities, but for their reputation. Boccaccio's book includes outstanding examples of evil women as well as of good ones, because he hoped that their lives would inspire his readers to behave virtuously. His first entry was Eve, the beauty born from Adam's rib and cursed 'with a woman's fickleness . . . foolishly, she thought that she was about to rise to greater heights'.[3]

Boccaccio inspired the first famous collection of women by a woman – Christine de Pizan's *Book of the City of Ladies*. The daughter of the French king's favourite mathematical astrologer, Christine de Pizan (1365–*c*.1430) burned with resentment at being denied the education she craved. One night, she was visited by three allegorical Ladies: Reason, Rectitude and Justice. Why, she asked, have men always treated women so badly? Because, answered Lady Reason, they secretly fear women's superiority. As proof of women's capabilities, this spiritual trio helped Christine de Pizan to construct (in her vivid imagination) an ideal city from the finest materials, where women could live together until eternity, honoured for their contributions to civilisation. She lived at a time when military valour and Christian virtue were highly valued, and so Christine de Pizan placed Amazonian warriors and the Virgin Mary inside her female utopia.[4]

As interest in intellectual women grew, Minerva and Christine de Pizan themselves became biographical collectors' items, set side by side with other heroines. When Eberti compiled his *Open Cabinet*, learned women were starting to become renowned for their scientific prowess as well as for their more conventional literary achievements. Well-intentioned men of science began to publish accounts celebrating female accomplishments, such as Lalande's tribute to women astronomers in the early nineteenth century. Since then, collections devoted to women's lives in science have appeared in increasing numbers and have expanded in size.[5]

But do such collections represent a more enlightened attitude? Many objections are summed up by the title of an early twentieth-century classic, reprinted in America as recently as 1991: *Woman in Science: with an introductory chapter on women's long struggle for things of the mind*. The author was a Catholic priest writing under the pseudonym H. J. Mozans. By using the singular *Woman*, he eliminated female individuality and set women apart as something special. His Darwinian subtitle reinforced this notion that women belong almost to a distinct species, like the masculine Minerva or those allegorical goddesses who represented the sciences but were unable to practise them. Encyclopaedias of women's lives have proliferated during the last few decades, but although they are – like Judy Chicago's *Dinner Party* – compiled with egalitarian zeal, they do perpetuate this idea that women should be celebrated separately from men.

Pandora's Breeches might qualify as one of these compilations, but it has been written with different ends in view. Rather than creating new female heroines, it has undermined conventional views of the past by attacking the very concept of heroism in science. This book has presented new interpretations of scientific men as well as of scientific women. We need to rewrite science's past by eliminating romanticised tales of lone geniuses and their glorious discoveries. Science is a collaborative project whose successes – and failures – can only be appreciated by understanding how scientific technology has permeated the whole of society. In revised versions of science's past, women have vital roles to play. Their contributions were often different from men's, but that does not mean that they were less important.

Like science itself, historical research is also a cooperative endeavour. *Pandora's Breeches* is the collection of one particular author who set up her own criteria to determine who should enter her cabinet and be put on show. She picked an international range of examples who would illustrate various ways in which women have contributed to the growth of science. There are many more stories to be told, many other forgotten scientific workers of the past – men as well as women – waiting to be revealed and reappraised. Anyone who reinterprets the past need never contemplate the collector's nightmare of a completed set.

Notes

Pandora/Eve/Minerva

Prologue

1. Dalton and Hamer, *Provincial Token-Coinage*, vol. 1, numbers 688, 787, 827, 829, 839–41, 1110, 1121. Several 'Guy Vaux' caricatures implicated Thomas Paine, and Richard Newton satirised George III holding up 'The Combustible Breeches'.
2. Wollstonecraft, *Vindication*, p. 317, 89, 88.
3. Wollstonecraft, *Vindication*, p. 7 (couplet from the 1801 *Anti-Jacobin Review and Magazine*).
4. This account is indebted to Barrell, *Birth of Pandora*, pp. 145–220 (Thomas Cooke, translator of Hesiod, quoted on p. 148). Rogers, *Troublesome Helpmate*.
5. Quoted in Baldick, *In Frankenstein's Shadow*, p. 22 (on the French Revolution).
6. Wollstonecraft, *Vindication*, p. 119.
7. Barry, *Works*, vol. 2, p. 383 (from his 1783 *Account* of the sixth picture in his series for the Royal Society of Arts).
8. Clarke, *Johnson's Women*, p. 27 (from Piozzi's *Anecdotes*).
9. Quoted in Bell, "'All clear sunshine'", p. 524 (letter to Polly Stevenson of 1 May 1760).
10. *Gentleman's Magazine* 8 (1738), 591 (reprinted from *Universal Spectator* of 25 November 1738).

1. Women/Science

1. Woolf, *A Room of One's Own*, pp. 48–53.
2. Many books and articles now discuss women in science. In addition to bibliographic encyclopaedias, valuable guides to the literature and issues include: Abir-Am and Outram, *Uneasy Careers*; Benjamin, *Science and Sensibility* and *Question of Identity*; Gates and Shteir, *Natural Eloquence*; Jordanova, "Gender and

237

historiography"; Hunter and Hutton, *Women, Science and Medicine*; Kohlstedt and Longino, *Women, Gender, and Science (Osiris* 12 (1997)); Pycior, Slack and Abir-Am, *Creative Couples*; Rossiter, *Women Scientists in America to 1948*; Schiebinger, *The Mind Has No Sex?* and *Has Feminism Changed Science?* For a comparable study of women historians, see Smith, *Gender of History*.

3. Makin, *Essay to Revive Education*, p. 35. Vickery, *Gentleman's Daughter*, pp. 127–60, and Hutton, 'Anne Conway, Margaret Cavendish', pp. 231–2.

4. Martensen, 'Transformation of Eve'; Schiebinger, *The Mind Has No Sex?*, pp. 160–244.

5. Miller, *Humours of Oxford*, p. 82. Discussions include Mullan, 'Gendered knowledge, gendered minds', and Reynolds, *Learned Lady*, pp. 373–419.

6. Centlivre, *Basset-Table*, p. 30.

7. Vickery, 'Golden age'. Grant, *Margaret the First*, pp. 151–2 (this frontispiece illustrated a book of stories, not one of her weightier works on natural philosophy). Cahn, *Industry of Devotion*; Laurence, *Women in England*.

8. Kargon, *Atomism*, pp. 66–76; Jones, *Glorious Fame*; Schiebinger, *The Mind Has No Sex?*, pp. 47–59.

9. Hunter, 'Sisters of the Royal Society'; Shapiro, 'Princess Elizabeth and Descartes'; Walters, 'Conversation pieces'.

10. Hunter, 'Sisters of the Royal Society', and Harris, 'Living in the neighbourhood'.

11. Aubrey, *Brief Lives*, p. 33 (Mary Sidney, sister of Philip). Hannay, '"How I these studies prize"'.

12. Guest, *Small Change*, and Goodman, *Republic of Letters*. Fara, *Newton*, pp. 136–7.

13. Gascoigne, *Banks*, pp. 23–6.

14. Simon, 'Mineralogy', quotation on p. 132.

15. Harkness, 'Managing an experimental household'.

16. Uglow, *Lunar Men*, pp. 313, 383.

17. Iliffe and Willmoth, 'Astronomy and the domestic sphere', pp. 244–57.

18. Lalande, *Ladies Astronomy*. Connor, 'Mme Lepaute'; Poirier, *Histoire des femmes de science*, pp. 155–64.

19. Fara, *Newton*, pp. 89–97.

20. Uglow, *Lunar Men*, p. 86.

21. Uglow, *Lunar Men*, pp. 313, 320.

22. Gould, 'Invisible woman', pp. 27–8; Richmond, 'Early history of genetics', pp. 87–90. Henry, *Prisoner of History*, pp. 121–4.

23. Tomaselli, 'Collecting women'. For eighteenth-century women's biographical collections, see Guest, *Small Change*, pp. 49–50, 64–9, 168–72. Important examples of this genre for scientific women include: Alic, *Hypatia's Heritage*; Mozans, *Woman in Science* (note the interesting singular!); Osen, *Women in Mathematics*; Rebière, *Les femmes dans la science*. Encyclopaedic reference books are their more scholarly successors. Particularly useful ones include: Creese, *Ladies in the Laboratory*; Kass-Simon and Farnes, *Women of Science;* Ogilvie and Harvey, *Biographical Dictionary*; Poirier, *Histoire des femmes de science*; Weisbard, *History of Women*.

24. Neeley, *Somerville*, pp. 11–44, 199–239. Tabor, *Pioneer Women*, exemplifies this bifurcated approach.

25. From *Biography for Beginners* by E. C. Bentley (the C stands for Clerihew). Wagner-Martin, *Telling Women's Lives* (for Woolf and Eliot, see pp. 11, 32–8); Backscheider,

Reflections on Biography, pp. 127–62 and Heilbrun, *Writing a Woman's Life*. See also Richards, *Angles of Reflection*.

26. Pycior, 'Pierre Curie and his "eminent collaborator"'.

27. See Shapin, *Scientific Revolution*, pp. 167–211 for a bibliographic essay on the enormous literature. Cunningham and Williams, 'De-centring "the big picture"'.

28. Nye, 'Masculine "fields of honor"'.

29. Carroll, 'The politics of "originality"'. Pumfrey, 'Who did the work?'; Shapin, 'House of experiment'; Secord, 'Science in the pub'. See also Anderson and Zinsser, *History of Their Own*.

30. Cooter and Pumfrey, 'Separate spheres and public places'.

31. Charles Lyell quoted in Farrell, 'Gentlewomen of science', p. 153.

32. Fores's quotations from Jordanova, *Nature Displayed*, pp. 23–5.

33. Jordanova, *Nature Displayed*, pp. 21–47; Porter, 'Touch of danger'; Vickery, *Gentleman's Daughter*, pp. 90–110; Wilson, *Making of Man-Midwifery*.

34. Darwin, *Correspondence*, vol. 4 , p. 147 (letter to Emma of 27–8 May 1848). Browne, *Darwin*, especially pp. 40, 76–7.

35. Browne, *Darwin*, especially pp. 146–7, 359–62.

36. Farrell, 'Gentlewomen of science', especially pp. 153–92 (quotation from a letter to her sister of 6 January 1854).

37. Hartman, *Victorian Murderesses*.

2. Lady Philosophy/Francis Bacon

1. Joseph Glanvill, *Plus Ultra: The Progress and Advancement of Knowledge Since the Days of Aristotle* (1668). Hunter, *Science and Society*, pp. 8–31, pp. 194–7, quotation on p. 195 (John Beale). For a recent survey of relevant Baconian literature, see Iliffe, 'Masculine birth of time'.

2. Hunter, *Science and Society*, pp. 194–7.

3. Bennett, 'Mechanics' philosophy'.

4. Sprat, *History of the Royal-Society*, pp. 63, 72.

5. Cavendish, *Observations on Experimental Philosophy*, pp. 43–4. Mintz, 'Duchess of Newcastle's visit'. For different interpretations of Cavendish, see: Grant, *Margaret the First*; Harris, 'Living in the neighbourhood'; Hutton, 'Anne Conway, Margaret Cavendish'; Jones, *Glorious Fame*; Schiebinger *The Mind Has No Sex?*, pp. 25–6, 47–59.

6. Schiebinger, *The Mind Has No Sex?*, pp. 119–59 (picture reproduced on p. 142). Beretta, 'Source of western science'. Sprat, *History of the Royal-Society*, p. 327, 124. For female nature in this period, key studies include: Easlea, *Witch-Hunting* and *Fathering the Unthinkable*; Jordanova, *Sexual Visions* and *Nature Displayed*; Merchant, *Death of Nature* and 'Isis' consciousness'; Schiebinger, *The Mind Has No Sex?* and *Nature's Body*.

7. Richardson, *Iconology*, vol. 1, pp. 61–6. This book was closely modelled on Cesare Ripa's *Iconologia*, first printed with figures in 1603.

8. Turnbull, vol. 2, pp. 437, 441 (letters of 20 and 29 June 1686).

9. 'To the *Royal Society*', lines 5–7 of Abraham Cowley's unpaginated poem in Sprat, *History of the Royal-Society*.

10. Boyle, *Works*, vol. 1, p. 310 (from *Some Considerations Touching Experimental Essays*, 1661).

11. Originally in Bacon's *Religious Meditations* ('Of heresies'). Sprat, *History of the Royal Society*, p. 129.

12. Joseph Glanvill quoted in Merchant, 'Isis' consciousness', p. 404. Rogers, *Troublesome Helpmate*.

13. Veldman and Luijten, *Dutch and Flemish Etchings: Martin van Heemskerck*, vol. 2, pp. 181–5. I am grateful to Anthony Snodgrass for this translation of the (contemporary) Latin, which was almost certainly specially composed by Hadrianus Junius, but does not entirely match the picture: 'Just as the mother bird leads its feathered offspring out from their early nests into the sky with a fixed melody (rhythm?), urging them to follow and rise up on their little feathers; so Nature, the renewer of everything, providently leads the human race from its soft cradles and the pregnant womb of the parent towards toils and hard labour.' Some versions have this French caption: '*Nature au monde met l'homme pour travailler, / Ainsi qu'elle y produit tout oyseau à voller*' (Nature puts mankind into the world to work, just as she makes every bird to fly).

14. Glanvill, *Vanity of Dogmatizing*, pp. 118, 135. Golinski, 'Care of the self'.

15. Bacon, *Works*, vol. 4, p. 296 (from 'Of the dignity and advancement of learning'). For Bacon, science and gender, see: Hutton, 'Riddle of the sphinx'; Iliffe, 'Masculine birth of time', pp. 441–5; Keller, *Reflections on Gender and Science*, pp. 33–65 and 'Secrets of God, nature and life'.

16. Bacon, *Works*, vol. 4, pp. 20–1 (preface to *The Great Instauration*). Farrington, *Philosophy of Bacon*, p. 62 (from 'The masculine birth of time').

17. Farrington, *Philosophy of Bacon*, p. 72 (from 'The masculine birth of time').

18. Isaac Barrow quoted in Easlea, *Witch-Hunting*, p. 246. Edmond Halley, letter to Newton of 1687, Turnbull *et al.*, *Correspondence of Newton*, vol. 2, p. 474.

19. Oldenburg, *Correspondence*, vol. 4, p. 77 (letter to Jean-Baptiste Denis of 23 December 1667 about blood transfusion). John Webster and John Evelyn quoted in Easlea, *Witch-Hunting*, p. 246.

20. 1767 letter quoted in Uglow, *Lunar Men*, p. 143.

21. Farington, *Philosophy of Bacon*, p. 129 (from 'The refutation of philosophies'). Davy, *Works*, vol. 8, pp. 175–6 (1807 lecture on chemistry). Crookes quoted in Easlea, *Fathering the Unthinkable*, p. 59.

22. Jordanova, *Sexual Visions*, pp. 87–110; Reay, *Watching Hannah*, pp. 46–50.

23. Willard Libby quoted in Easlea, *Fathering the Unthinkable*, p. 170. Geologist (on the 1980 Mount St Helens eruption) quoted in Merchant, 'Isis' consciousness', p. 405.

24. Two starting points for references to the literature are Kohlstedt and Longino, *Women, Gender, and Science*, and Schiebinger, *Has Feminism Changed Science?*

In the Shadows of Giants

1. Merton, *Shoulders of Giants*.

2. Du Châtelet, *Institutions de physique*, p. 6 ('*Nous nous élevons à la connaissance de la vérité, comme ces Géans qui escaladoient les Cieux en montant sur les épaules les uns des autres*'). For other interpretations of this image, see Harth, *Cartesian Women*, p. 198 n.48, and Schiebinger, *The Mind Has No Sex?*, pp. 130–31.

3. Elisabeth of Bohemia/René Descartes

1. C. Descartes, 'Mort de Descartes', p. 131. Image originally drawn by Bernard Picart for a dissertation by Brillon de Jouy, and later plagiarised by an anonymous artist. *Nouvelles de la République des Lettres*, August 1707, pp. 232–5 (by Jacques Bernard). Discussed in Duportal, *Picart*, p. 369, and Saxl, 'Veritas', pp. 218–22.

2. My major sources of information on Christina are Åkerman, *Queen Christina*, pp. 44–69, 103–77, Brummer, 'Minerva of the north', and Vrooman, *Descartes*, pp. 212–48. For the importance of royal patronage, see Ashworth, 'Habsburg circle', Biagioli, *Galileo Courtier*, and Cormack, 'Twisting the lion's tail'.

3. C. Descartes, 'Mort de Descartes', pp. 131–2; see also Harth, *Cartesian Women*, pp. 95–8, and Sorbière, *Sorberiana*, p. 78 (on Descartes sleeping with naked nature). Gaukroger, *Descartes*, pp. 412–17.

4. Gaukroger, *Descartes*, pp. 1–14; this 'intellectual biography' is my major source of information about Descartes.

5. Bordo, 'Cartesian masculisination of thought'; Harth, *Cartesian Women*, pp. 235–9.

6. The only substantial book-length biography of Elisabeth in English is Godfrey, *Sister for Prince Rupert*, which is almost a hundred years old. Other biographical accounts include Adam, *Vie et œuvres de Descartes*, pp. 401–31, Cohen, *Écrivains*, pp. 603–51, and Zedler, 'Three princesses', pp. 29–33. In my account of the relationship between Elisabeth and Descartes I have relied mainly on Harth, *Cartesian Women*, pp. 62–78, Néel, *Descartes et Elisabeth*, and Zedler, 'Three princesses', pp. 33–43; other analyses include: Mattern, 'Descartes's correspondence with Elizabeth'; Shapiro, 'Princess Elisabeth and Descartes'; and Vrooman, *Descartes*, pp. 167–211. The correspondence between Descartes, Elisabeth, Christina and Chanut is conveniently gathered together in one volume, Descartes, *Lettres sur la morale*; this is an expanded version, in modernised French, of the publication in 1879 of Elisabeth's letters after missing ones were rediscovered and transcribed by Alexandre Foucher de Careil. All translations from this source are my own. The 1989 edition of this correspondence by Jean-Marie and Michelle Beyssade is not easily obtainable in England.

7. Godfrey, *Sister for Prince Rupert*, pp. 74–5. Honthorst taught Elisabeth and her younger sister Louise, whose own pictures are sometimes attributed to him. It seems possible, therefore, that Louise may have painted this portrait. In 1714, when Sophie's son George came to the English throne, the 'Hanover pearls' came to England.

8. Harth, *Cartesian Women*, p. 64. Descartes, *Lettres sur la morale*, p. 5 (21 May 1643) ('*un corps si semblable à ceux que les peintres donnent aux anges*').

9. Sorbière, *Sorberiana*, pp. 85–6.

10. Descartes, *Lettres sur la morale*, p. 5 (21 May 1643).

11. By Claude Clerselier, in 1657 (letters 3–31, plus others).

12. Samuel Sorbière dedicated his translation of Gassendi, and Constantijn Huyghens his own volume of poems. Rang, '"An exceptional mind"'.

13. Israel, *Radical Enlightenment*, especially pp. 23–58. Descartes, *Lettres sur la morale*, p. 65 (letter from Elisabeth of 16 August 1645). Rang, '"An exceptional mind"'.

14. Harth, *Cartesian Women*, pp. 63–78. Descartes, *Lettres sur la morale*, p. 114 (letter from Descartes to Elisabeth of 3 November 1645).

15. Gaukroger, *Descartes*, pp. 354–83, especially pp. 361–4.

16. Descartes, *Principles of Philosophy*, pp. xiv–xvi (originally written in Latin).

17. Descartes, *Lettres sur la morale*, pp. 34–9 (Elisabeth's letter of 1 August 1644 ['*Je crains que vous rétracterez, avec justice, l'opinion que vous eûtes de ma comprehension, quand vous saurel que je n'entends pas comment . . .*'] and Descartes's reply later in August ['*Je supplie très humblement Votre Altesse de me pardonneer, si je n'écris rien ice que fort confusément. Je n'ai point encore le livre . . . et je suis en un voyage continu*']).

18. Descartes, *Lettres sur la morale*, pp. 3–9 (Elisabeth's letter of 16 May 1643, and Descartes's reply of 21 May 1643).

19. Descartes, *Lettres sur la morale*, pp. 10–28 (Elisabeth's letter of 20 June 1643, Descartes's reply of 28 June 1643, her reply of 1 July 1643).

20. This definition of wisdom is in the 1647 preface to the French translation: Descartes, *Principles of Philosophy*, p. xvii. Shapin, 'Descartes the doctor' (quotation on p. 136).

21. Descartes, *Lettres sur la morale*, pp. 44–5 (letter from Elisabeth to Descartes of 24 May 1645: '*le corps . . . se ressent très facilement des afflictions de l'âme*'); p. 190 (letter from Descartes to Elisabeth of July 1647: '*l'âme, qui a sans doute beaucoup de force sur le corps, ainsi que montrent les grands changement* [sic] *que la colère, la crainte et les autres passions excitent en lui*'); p. 159 (letter from Descartes to Elisabeth of November 1646: '*lorsque l'esprit est plein de joie, cela sert beaucoup à faire que le corps se porte mieux*'). Gaukroger, pp. 384–417, especially p. 388.

22. Descartes, *Lettres sur la morale*, pp. 57–67 (Descartes's letters of 21 July and 4 August 1645, and Elisabeth's reply of 16 August 1645.

23. Golinski, 'Care of the self' (quotation on p. 131); Shapiro, 'Princess Elisabeth and Descartes', pp. 508–17.

24. Néel, *Descartes et Elisabeth*, pp. 60–102.

25. Descartes, *Lettres sur la morale*, p. 87 (Elisabeth's letter of 13 September 1645) and p. 115 (Descartes's letter of 3 November 1645).

26. Gaukroger, *Descartes*, pp. 384–405.

27. Descartes, *Lettres sur la morale*, pp. 163–6 (Elisabeth's letter of 29 November 1646) and p. 238 (Descartes's letter of 1 November 1646 to Chanut: '*surpasser de beaucoup en érudition et en vertu les autres hommes*'). Gaukroger, *Descartes*, pp. 412–16.

28. Descartes, *Lettres sur la morale*, pp. 221–2 (Descartes's letter to Elisabeth of 9 October 1649).

29. Descartes, *Lettres sur la morale*, pp. 223–4 (Elisabeth's letter of 4 December 1649).

30. Gaukroger, *Descartes*, pp. 1–6. Vrooman, *Descartes*, pp. 249–61 (Descartes quoted on p. 257, Voltaire quoted on p. 258).

31. Åkerman, *Queen Christina*, pp. 44–69.

32. Sophia, *Memoirs*, pp. 41–3, 251–2; Godfrey, *Sister for Prince Rupert*, pp. 223–353.

33. Harth, *Cartesian Women*, especially pp. 66–7, 235–9.

4. Anne Conway / Gottfried Leibniz

1. Aiton, *Leibniz*.

2. Zedler, 'Three princesses', pp. 43–51. Sophie, *Memoirs*.

3. Gerhardt, *Schriften*, vol. 3, p. 367: letter to Damaris Masham of 10 July 1705

('*cette grande Princesse . . . se plaisoit à estre informée de mes speculations, elle les appro-fondissoit même*').

4. Schiebinger, 'Winkelmann', p. 195 (letter to Sophie Charlotte of November 1697) and *The Mind Has No Sex?*, p. 98.

5. Zedler, 'Three princesses', pp. 51–9, quotation on p. 59.

6. Berteloni Meli, 'Caroline, Leibniz, and Clarke'.

7. Gerhardt, *Schriften*, vol. 3, pp. 336–75 (reference to Conway on p. 337). Frankel, 'Damaris Cudworth Masham'.

8. My major sources for this account of Conway and her philosophy are: Conway, *Principles of Philosophy*, pp. vii–xxxiii; Duran, 'Anne Viscountess Conway'; Merchant, 'Vitalism of Anne Conway'; Nicolson and Sutton, *Conway Letters*.

9. Gerhardt, *Schriften*, vol. 3, p. 176: letter to Thomas Burnett of 17 March 1696 ('*Il a esté ami particulier de Mad. la Comtesse de Kennaway, et il me conta l'histoire de cette Dame extraordinaire*'). Aiton, *Leibniz*, pp. 201–2.

10. Harris, 'Living in the neighbourhood' (quotation on p. 201).

11. More, *Antidote against Atheism*, p. 5 of unpaginated preface.

12. Nicolson and Hutton, *Conway Letters*, pp. 128–9 (letter from More to Conway of 7 January 1656).

13. Brusati, pp. 126, 201–17. Cust and Malloch, 'Portraits'.

14. Laurence, *Women in England*, pp. 108–80. Vickery, *Gentleman's Daughter*, pp. 127–60. Hutton, 'Anne Conway, Margaret Cavendish', pp. 227–32. Hunter, 'Sisters of the Royal Society'.

15. Popkin, 'Spiritualistic cosmologies'.

16. Nicolson and Hutton, *Conway Letters*, pp. 91 (10 December 1653), 105 (5 September 1654).

17. Conway, *Principles of Philosophy*, p. 58.

18. Martensen, 'Transformation of Eve' (quotations on p. 117).

19. Golinski, 'Care of the self', pp. 131–3 (quotation on p. 132).

20. Gerhardt, *Schriften*, vol. 3, p. 217: letter to Thomas Burnett of 24 August 1697 ('*les miens en philosophie approchent un peu davantage de ceux de feu Mad. la Contesse do Conway, et tiennent le milieu entre Platon et Democrite . . . tout estant plein de vie et de perception*').

21. Conway, *Principles of Philosophy*, p. 20. Merchant, 'Vitalism of Anne Conway', exaggerates Conway's influence.

22. Coudert, *Leibniz and the Kabbalah*, especially pp. 25–77.

23. Conway, *Principles of Philosophy*, p. 43. Hutton, 'Of physic and philosophy'.

24. Conway, *Principles of Philosophy*, p. 3.

5. *Emilie du Châtelet / Isaac Newton*

1. Voltaire quotations: letter to Frederick II of 15 October 1749: Voltaire, *Correspondence*, vol. 17, p. 188 ('*un grand homme qui n'avoit de défaut que d'être femme*'), and Edwards, *Divine Mistress*, p. 48. Important revisionist articles about du Châtelet's life and role include Terrall, 'Du Châtelet and the gendering of science', and also Zinsser, 'Du Châtelet: genius, gender, and intellectual authority' and 'Entrepreneur of the "Republic of Letters"'. For discussions of her ideas, see Iltis, 'Du Châtelet's metaphysics and mechanics', and Janik, 'Du Châtelet's

Institutions de physique'. See also Harth, *Cartesian Women*, pp. 189–213. For her influence on Voltaire, see Wade, *Voltaire and Madame du Châtelet*, especially pp. 3–47, and *Intellectual Development of Voltaire*, pp. 251–570.

2. Voltaire, *Élémens de Newton*, p. 3 ('*vaste & puissant Génie, / Minerve de la France*'). Walters, 'Allegorical engravings'.

3. For Newton's life and works, see Westfall, *Never at Rest*. For Pope and Newtonian myths, see Fara, *Newton*, pp. 39–58, pp. 192–230.

4. Voltaire, *Letters on England*, pp. 69, 71.

5. Brook Taylor (1718, to Rémond de Monmort), quoted in Fara, *Newton*, p. 128.

6. Letter of April 1736 to Francesco Algarotti: Besterman, *Lettres*, vol. 1, p. 111. Voltaire's preface to du Châtelet, *Principes mathématiques*, p. xi ('*C'étoit un avantage qu'elle eut sur Newton, d'unir à la profondeur de la Philosophie, le goût le plus vif & le plus délicat pour les Belles Lettres*'). For Newton in France, see Fara, *Newton*, pp. 126–54. Book-length biographies of du Châtelet include: Badinter, *Émilie, Émilie*; Edwards, *Divine Mistress*; Ehrman, *Mme du Châtelet*; see also Poirier, *Femmes de science*, pp. 235–59.

7. Du Châtelet quoted in Edwards, *Divine Mistress*, p. 1 (from a letter to Frederick the Great). Letter to Pierre le Cornier de Cideville of 15 March 1739: Besterman, *Lettres*, vol. 1, p. 346.

8. Zinsser, 'Du Châtelet: genius, gender, and intellectual authority'. Fara, 'Elizabeth Tollet'.

9. Du Châtelet quoted in Edwards, *Divine Mistress*, p. 1 (from a letter to Frederick the Great). Voltaire, *Élémens de Newton*, p. 3 ('*vaste et puissant Génie, / Minerve de la France*'). Walters, 'Allegorical engravings'.

10. Letter of May 1738 to Maupertuis: Besterman, *Lettres*, vol. 1, p. 232 ('*M'avez-vous reconnue dans l'estampe de la Luce?*'); see also letter of April 1736 to Algarotti: Besterman, *Lettres*, vol. 1, p. 111. My account is mainly based on Ehrman, *Mme du Châtelet*, Terrall, 'Du Châtelet and the gendering of science', and Zinsser, 'Du Châtelet: genius, gender, and intellectual authority'.

11. Mme du Deffand, quoted in Poirier, *Femmes de science*, p. 235. ('*Représentez-vous une femme grande et sèche . . . le visage aigu, le nez pointu . . . le teint noir, rouge, échauffé . . . Voilà la fugure de la belle Emilie, figure don't elle est si contente qu'elle n'épargne rien pour la faire valour: frisures, pompons, pierreries, verreries, tout est à profusion.*')

12. From *Discours sur le bonheur*, quoted in Ehrman, *Mme du Châtelet*, p. 80.

13. Kant, 'Observations on the beautiful and the sublime', p. 62 (1764).

14. Madame du Deffand quoted in Poirier, *Femmes de science*, p. 237 ('*Emilie travaille avec tant de soin à paraître ce qu'elle ne'est pas, qu'on ne sait plus ce qu'elle est en effet*').

15. Zinsser, 'Entrepreneur of the "Republic of Letters"', pp. 602–3.

16. Letter to Maupertuis of 21 June 1738: Besterman, *Lettres*, vol. 1, p. 236 ('*J'ai voulu essayer mes forces à l'abri de l'incognito*').

17. Voltaire's *Mémoires* quoted in Edwards, *Divine Mistress*, p. 86. Findlen, 'A forgotten Newtonian' (Algarotti quoted on p. 315) and Mazzotti, 'Maria Gaetana Agnesi'. See also Harth, *Cartesian Women*, pp. 199–205.

18. Besterman, *Lettres*, vol. 1, p. 255. Letter of September 1738 to Maupertuis ('*Son livre est frivole . . . il pourra réussir aux toilettes*').

19. Letter to Maupertuis of 18 July 1736: Besterman, *Lettres*, vol. 1, p. 121. Iliffe, '"Aplatisseur du monde"'.

20. Letter to Cideville quoted in Poirier, *Femmes de science*, p. 245 ('*Madame du Châtelet est dans tout cela mon guide et mon oracle*'). Wade, *Voltaire and du Châtelet*, pp. 34–40. Letter to Prince Frederick of January 1737, quoted in Janik, 'Searching for the metaphysics', p. 88. Voltaire, *Élémens de Newton*, pp. 9–10 ('*L'étude solide que vous avez faite de plusieurs nouvelles vérités & le fruit d'un travail respectable sont ce que j'offre au Public pour votre gloire*'), and pp. 7–8 ('*Puissai-je auprès de vous . . . Aux regards des Français montrer la Vérité*'). See Voltaire, *Correspondence*, vol. 5, pp. 207, 208, 230, 236 and 264 (letters written in August and September 1736).

21. Voltaire quotation from the 1748 *Epître dédicatoire*, quoted in Wade, *Voltaire and du Châtelet*, p. 37 ('*je m'instruisais avec vous. Mais vous avez pris depuis un vol que je ne peux plus suivre*'). For their comments on the book's scope, see du Châtelet, *Institutions de physique*, p. 7, and Voltaire, *Élémens de Newton*, p. 11; discussed in Janik, 'Searching for the metaphysics', p. 88.

22. The following account is based on Iltis, 'Du Châtelet's metaphysics', and Janik, 'Searching for the metaphysics'.

23. Neeley, *Mary Somerville*, p. 76.

24. Zinsser, 'Entrepreneur of the "Republic of Letters"', especially p. 608. The following account is closely based on Zinsser, 'Translating Newton's *Principia*'.

25. Sørenson, 'Méditation dans les ruines'.

26. *Journal Encyclopédique* 6 (September 1759), part 3, pp. 3–17, quotation on p. 5 ('*pour la rendre plus éclatante, en la faisant concourir avec le moment du triomphe de la philosophie*'). See Schaffer, 'Halley, Deslisle, and the making of the comet'.

27. Evans, 'Fraud and illusion'.

Domestic Science

1. Martin, *Young Gentlemen's and Ladies Philosophy*, vol. 1, p. 2.

2. Mullan, 'Gendered knowledge, gendered minds'; Myers, 'Science for women and children'; Walters, 'Conversation pieces'. Carroll, 'Politics of "originality"'.

3. Ferguson, *Young Gentleman and Lady's Astronomy*, p. 2, 45. Walters, 'Conversation pieces', pp. 132–3.

4. Withering Correspondence, Royal Society of Medicine, ff. 66–94, 145–6 (quotations from f. 67, letter of 30 November 1784). See entry by Patricia Fara and Anne Secord in Ogilvie, *Women in Science*.

5. Wright and Knowles, both quoted on p. 354 of Uglow, *Lunar Men*.

6. Tollet, *Poems*, pp. 25–6 (from 'To my Brother at St John's College in Cambridge'). Fara, 'Elizabeth Tollet'.

7. Iliffe and Willmoth, 'Astronomy and the domestic sphere', pp. 240–1. Bell, '"All clear sunshine"'; Heilbron, 'Franklin'. Costa, '*Ladies' Diary*'.

6. *Jane Dee/John Dee*

1. Martin, *Young Gentlemen's and Ladies Philosophy*, vol. 1, p. 316. Sidrophel was the conjurer in Samuel Butler's *Hudibras* (1663).

2. http://www.johndee.org/DEE.html, 8 February 2002. French, *Dee*, pp. 4–19.

3. Quoted in French, *Dee*, p. 110 (from *A True & Faithful Relation of what passed for many Yeers Between Dr: John Dee . . . and Some Spirits*). See Harkness, *Dee's Conversations with Angels*.

4. Dee, 'Preface', cuts Aj and Aij. The most reliable general discussions of John Dee are French, *Dee*, and Clulee, *Dee's Natural Philosophy*, although they barely mention Jane Fromond / Dee. For reconstructing Jane Dee's life and experiences, this chapter relies heavily on Harkness, 'Managing an experimental household', a splendid article to which I am much indebted.

5. Harris, 'Living in the neighbourhood of science'. Hunter, 'Sisters of the Royal Society'.

6. Fenton, *Diaries*, p. 44 (6 May 1582, deleted together with words in Kelly's hand). John Dee's mother was also called Jane.

7. Cahn, *Industry of Devotion* (Heinrich Bullinger, *The Christian State of Matrimony*, quoted p. 89), and Laurence, *Women in England*.

8. Wilson, *Midwifery*.

9. Fenton, *Diaries*, pp. 174–5 (21 March 1585).

10. Fenton, *Diaries*, p. 36 (19 March 1582).

11. Fenton, *Diaries*, p. 218 (18 April 1587).

12. Fenton, *Diaries*, pp. 218 (18 April 1587), 223 (21 May 1587), 233 (28 February 1588).

13. *Spectator*, 12 March 1711.

14. Quoted from *Poor Richard's Almanac* in Heilbron, 'Franklin', p. 198.

7. *Elisabeth Hevelius / Johannes Hevelius*

1. Edmond Halley, quoted in Winkler and van Helden, 'Hevelius and visual language', p. 97 n.2. My major sources for Elisabetha Hevelius's biography are Cook, 'Johann and Elizabeth Hevelius', and MacPike, *Hevelius, Flamsteed and Halley* (account of their betrothal reproduced on pp. 4–5). In his *Machina Cælistis*, there are three pictures of them observing together: facing pp. 222 and 254, and after p. 420.

2. Biagioli, *Galileo Courtier*; Ashworth, 'Habsburg circle'.

3. Latin preface translated in MacPike, *Hevelius, Flamsteed and Halley*, pp. 122–4 (quotations on p. 123).

4. My major source for discussing Johannes Hevelius is Winkler and van Helden, 'Hevelius and visual language'.

5. Quoted in Cook, 'Johann and Elizabeth Hevelius', p. 10 ('*conjugam meam charissimam*'; '*Quippe ad observationis Mulieres aeque at Viri idoneæ*'). For tables of labelled observations, see Cook, *Halley*, pp. 94–7.

6. MacPike, *Hevelius, Flamsteed and Halley*, pp. 75–102, quotation on p. 75; Cook, *Halley*, pp. 89–127.

7. Thomas Hearne, quoted in Cook, *Halley*, p. 102.

8. The following discussion closely follows Schiebinger, 'Maria Winkelmann' and *The Mind Has No Sex?*, pp. 66–101. For details of Winkelmann's life, both of these rely on Vignoles, 'Eloge de Madame Kirch', especially pp. 168–83.

9. Iliffe and Willmoth, 'Astronomy and the domestic sphere', pp. 244–56.

10. Celsisus quoted in Mozans, *Woman in Science*, p. 174.

11. Millburn, *Adams of Fleet Street*.

12. Vignoles, 'Eloge de Madame Kirch', pp. 183 (*'on doit reconnoitre, en general, qu'il n'y a aucune espéce de Science, dont les Femmes ne soient capable'*) and 181 (*'& auroient voulu la réduire à sa quenouille, & à son fuseau'*).

13. Quoted in Schiebinger, *The Mind Has No Sex?*, p. 87.

14. Klopp, *Werke von Leibniz*, vol. 9, p. 295 (letter of January 1709: *'Je ne crois pas que cette femme trouve facilement sa pareille dans la science où elle excelle . . . Elle observe comme pourroit faire le meilleur observateur'*).

15. Johann Jablonski, quoted in Schiebinger, *The Mind Has No Sex?*, p. 92; Schiebinger, 'Maria Kirch'.

16. Klopp, *Werke von Leibniz*, vol. 9, p. 295 (letter of January 1709: *'une femme des plus savantes, qui peut passer pour une rareté . . . une des meilleures raretés de Berlin'*).

17. Quoted in Schiebinger, *The Mind Has No Sex?*, p. 80 (from Eberti's *Eröffnetes Cabinet deß gelehrten Frauen-Zimmers*). For Cunitz's life, see Vignoles, 'Eloge de Madame Kirch', pp. 163–8 (references to Eberti on p. 155, 165). For a brief discussion of Eberti and his *Cabinet*, see *Daphnis* 18 (1989), pp. 345–6.

8. Caroline Herschel / William Herschel

1. South, 'An address', quotations from pp. 410–11. I have relied heavily on two book-length biographies for Caroline: Herschel, *Memoir*, and Lubbock, *Herschel Chronicle*; however, I have only given page references for quotations. Major articles include Iliffe and Willmoth, 'Astronomy and the domestic sphere', especially pp. 257–62, and Ogilvie, 'Caroline Herschel's contributions'. For the Herschels' early years in Bath, see Turner, *Science and Music*. My two major sources for William Herschel's astronomy are Bennett, 'Power of penetrating', and Schaffer, 'Herschel in Bedlam'. My account is not strictly chronological, and leaves out some details: for instance, it does not describe how the Herschels moved between several houses in Bath. Caroline Herschel's autobiographies have never been published in their entirety, but substantial sections are reproduced in her biographies. I am grateful to the Humanities Research Center of the University of Texas at Austin for allowing me to read microfilm copies of her original manuscripts, M0675–7.

2. Herschel, *Memoir*, p. 227. Mary Somerville was awarded honorary membership at the same time.

3. From 'On first looking into Chapman's Homer'.

4. Lalande, *Ladies' Astronomy*, pp. x–xii, 93 (originally published in 1802). Barrett, *Diary and Letters*, vol. 2, p. 167 (21 August 1786).

5. Wallington, 'Physical and intellectual capacities', p. 557.

6. Herschel, *Memoir*, pp. ix, 37–8.

7. Barrett, *Diary and Letters*, vol. 4, pp. 25–6 (letter to Fanny Burney of 28 September 1797) and vol. 2, p. 409 (21 August 1787). Herschel, *Memoir*, p. 142 (John Herschel's wife Mary). Tabor, *Pioneer Women*, vol. 4, p. 6.

8. Herschel, *Memoir*, pp. 21, 7.

9. Lubbock, *Chronicle*: undated letter (probably January 1761) to his brother Jacob quoted on p. 17, and p. 45.

10. John Herschel's letter to R Wolf of 1866, quoted in Lubbock, *Chronicle*, p. 58.
11. Herschel, *Memoir*, p. 33.
12. Lubbock, *Chronicle*, p. 56.
13. Lubbock, *Chronicle*, p. 89.
14. Lubbock, *Chronicle*, p. 57.
15. Lubbock, *Chronicle*, pp. 78, 115 (letter of 3 June 1782).
16. Herschel, *Memoir*, p. 37.
17. Herschel, *Memoir*, p. 38. Letter to John Herschel of 1 February 1826: British Library Egerton Manuscripts 3761, f. 45r.
18. Lubbock, *Chronicle*, pp. 136, 138.
19. Lubbock, *Chronicle*, p. 150. For her telescopes, see Bullard, 'My small Newtonian sweeper', and Hoskins, 'Caroline Herschel's comet sweepers'. Letter to John Herschel of 1 February 1826: British Library Egerton Manuscripts 3761, f. 44r.
20. Faujas de Saint Fond, *Journey through England*, vol. 1, pp. 59–77 (quotation on p. 69).
21. Herschel, *Memoir*, p. 309.
22. Herschel, 'Account of the discovery of a comet', p. 1. Lubbock, *Chronicle*, pp. 153–6 (quotation from Alexander Aubert on p. 155).
23. Lubbock, *Chronicle*, p. 172.
24. British Library Egerton Manuscripts 3761, f. 26r (letter to Lady Herschel of 17 October 1824). Barrett, *Diary and Letters*, vol. 3, p. 45 (3 October 1788).
25. Letter to Edward Piggott of 6 December 1793 quoted in Hoskin, 'Caroline Herschel's comet sweepers', p. 31. Letter to Joseph Banks of 8 November 1795 quoted in Herschel, *Memoir*, p. 94.
26. Seymour, *Shelley*, pp. 27, 83.
27. Letter of 18 December 1801: British Library Additional Manuscripts 37203, f. 9r ('*puisque c'est actuellement la chose qui interesse le plus les astronomes. nous en attendons quelques unes de vous*'). Connor, 'Mme Lepaute'. Schaffer, 'Halley, Delisle, and the making of the comet'. Sir Harry Englefield and Joseph de Lalande: Lubbock, *Chronicle*, pp. 247, 251. Lalande, *Ladies' Astronomy*, pp. xi–xii.
28. Iliffe and Willmoth, 'Astronomy and the domestic sphere'.
29. Herschel, *Memoir*, p. 120.
30. Herschel, *Memoir*, pp. 126, 133.
31. British Library Egerton Manuscripts 3761, ff. 5r and 13v (letters of 12 November 1822 and 27 February 1823).
32. Herschel, *Memoir*, p. 231 (letter to John Herschel of 21 August 1828).
33. Letter of September 1798 to Nevil Maskelyne from Herschel, *Memoir*, p. 96.
34. Herschel, *Memoir*, p. ix; see also British Library Egerton Manuscripts 3761, ff. 31v–32r (letter to John Herschel of 14 January 1825).
35. Letter to Lalande of 3 March 1794: British Library Additional Manuscripts 37203, f. 6r. Letter to Lalande of 3 March 1794: British Library Additional Manuscripts 37203, f. 6r.
36. Letters to John Herschel of 24 June and 14 July 1823: British Library Egerton Manuscripts 3761, ff. 16v and 18r. Letter from Alexander Aubert to William Herschel of January 1822, quoted in Schaffer, 'Herschel in Bedlam', p. 211.
37. Wollstonecraft, *Vindication*, p. 179.
38. Letter of January 1800 to Nevil Maskelyne quoted in Herschel, *Memoir*, p. 102.

Letter to John Herschel of 1 December 1839: British Library Egerton Manuscripts 3762, f. 48r.

39. Montagu, *Letters*, vol. 3, p. 22, and vol. 2, p. 5. British Library Additional Manuscripts 37203, f. 9 (letter of 18 December 1801).
40. Anna Knipping, quoted in Herschel, *Memoir*, p. 346.

9. Marie Paulze Lavoisier / Antoine Lavoisier

1. My sources for analysing this portrait are Vidal, 'David among the moderns', and Beretta, *Imaging a Career in Science*, pp. 25–42. See also Brookner, *David*, p. 88; de Nanteuil, *David*, pp. 66–9; Poirier, *Lavoisier*, pp. 1–3; Schnapper, *David*, p. 84. The major biographies of Antoine Lavoisier that I have consulted, which also include substantial information about Paulze Lavoisier, are: Donovan, *Lavoisier*; McKie, *Lavoisier*; Poirier, *Lavoisier*. My main sources for this account of Paulze Lavoisier are Duveen, 'Madame Lavoisier' and Poirier, *Lavoisier*, especially pp. 94–6, 370–411. See also: Grimaux, *Lavoisier*, pp. 35–45, 330–6, 381–4; Hoffman, 'Mme. Lavoisier'; Poirier, *Femmes de science*, pp. 281–325; Smeaton, 'Monsieur and Madame Lavoisier' and 'Madame Lavoisier and du Pont de Nemours'.
2. Jean-François Ducis, quoted in Schnapper, *David*, p. 84 ('*Pour Lavoisier, soumis à vos lois / Vous remplissez les deux emplois / Et de muse et de sécretaire*').
3. Goodman, *Republic of Letters* (quotation on p. 7).
4. Vidal, 'David among the moderns'.
5. Outram, 'Before objectivity'. Rayner-Canham and Frenetter, 'Some French women chemists', pp. 176–7. Poirier, *Femmes de science*, pp. 263–80.
6. Du Pont de Nemours quoted in Poirier, *Lavoisier*, pp. 394 (23 October 1794), 126 (April 1815).
7. Rayner-Canham and Frenette, 'French women chemists', pp. 176–7; Combremont, *Necker de Saussure*, p. 58.
8. Gouverneur Morris quoted in Duveen, 'Madame Lavoisier', p. 17 (8 June 1789 to 7 November 1791).
9. Outram, 'Before objectivity', Franklin quoted on p. 23.
10. Letter of 1777, quoted in Duveen, 'Madame Lavoisier', p. 16 ('*me faire décliner et conjurer pour me faire plaisir et me rendre digne de mon mari*').
11. Poirier, *Lavoisier*, pp. 126–7 (François de Frénilly).
12. Young, *Travels in France*, p. 72.
13. Grimaux, *Lavoisier*, pp. 44–5 ('*C'était pour lui un jour de bonheur; quelques amis éclairés, quelques jeunes gens . . . se réunissaient dès le matin dans le laboratoire; c'était là que l'on déjeunait, que l'on dissertait, que l'on créait cette théorie qui a immortalisé son auteur*'). For the translation of the ambiguous '*on*' as 'we', see Vidal, 'David among the moderns', p. 613.
14. Young, *Travels in France*, p. 72. Beretta, 'Chemical imagery'. Duveen, 'Madame Lavoisier', pp. 17–19.
15. Beretta, *Imaging a Career in Science*, pp. 43–7.
16. Scheler, 'Deux lettres', p. 125 (letter to Guyon de Morveau of 16 September 1788) ('*deux minutes plus tard nous étions six personnes victimes*'). Paper quoted in Poirier, *Lavoisier*, p. 225.

17. Quoted in Grimaux, *Lavoisier*, p. 266 ('*que ces sangsues publiques soient arrêtées, et que, si leurs comptes ne sont pas rendus dans un mois, la Convention les livre au glaive de la loi*').
18. Joseph Lagrange quoted in Poirier, *Lavoisier*, p. 382.
19. Quoted in Poirier, *Lavoisier*, p. 357.
20. Quoted in Poirier, *Lavoisier*, p. 391.
21. Quoted in Poirier, *Lavoisier*, p. 393.
22. Bensaude-Vincent, 'Between history and memory', pp. 493–9; Beretta, *Imaging a Career in Science*, pp. 53–9. Smeaton, 'Madame Lavoisier and the publication of Lavoisier's *Mémoires de chimie*', (J. C. Delametherie quoted on p. 28). The preface is reproduced in Grimaux, *Lavoisier*, pp. 332–3.
23. Quoted in Duveen, 'Madame Lavoisier', p. 21.
24. Duveen, 'Madame Lavoisier', pp. 24, 26 (letters to his daughter of 24 October 1806 and 24 October 1810).
25. Guizot, 'Madame de Rumford', quotation on p. 83 ('*Il faut avoir vécu sous la machine pneumatique pour sentir tout le charme de respirer*').

Under Science's Banner

1. Edgeworth, *Letters for Literary Ladies*, p. 3.
2. Uglow, *Lunar Men*, pp. 181–8, Day quoted on p. 185. See also Douthwaite, *Wild Girl*, pp. 138–41.
3. Uglow, *Lunar Men*.
4. Letters to Withering of 5 December 1787 and 30 November 1784: Withering letters, Royal Society of Medicine, London, ff. 92 v, 66r, 67v. See entry by Patricia Fara and Anne Secord in Ogilvie, *Women in Science*, vol. 2, p. 1404.
5. Darwin, *Plan for Female Education*, pp. 40–5. See King-Hele, *Darwin*.
6. Darwin, *The Botanic Garden*, p. v. Danchin, 'Darwin's scientific and poetic purpose', Walpole quoted on p. 135.
7. Edgeworth, *Letters for Literary Ladies*, pp. 20–1.
8. Shelley, *Frankenstein*, p. 3 (by Percy Shelley).

10. Priscilla Wakefield / Carl Linnaeus

1. Wordsworth (*Grasmere Journal*, 1800) quoted on p. 21 of Shteir, *Cultivating Women*, the major work on botanical women for this period.
2. Bermingham, *Learning to Draw*, pp. 202–24, *Spectator* quoted on p. 204.
3. *Gentleman's Magazine* 71:1 (1801), 199–200.
4. This hand-written poem is on the endpapers of Joseph Banks's copy of Darwin's translation of Linnaeus's *A System of Vegetables* (British Library press mark 447.c.19). Ann Shteir suggests that the author was Anna Seward, which seems very likely. Extracts from ff. 2, 3, 5.
5. Thornton, *The Temple of Flora*, pp. 48–9.
6. By far the best biography of Linnaeus, on which this account is based, is Koerner, *Linnaeus*.
7. Koerner, 'Women and utility', Linnaeus quoted on p. 245 (a speech of 1772).

8. Amongst the substantial literature, I have drawn particularly from Bewell, '"Jacobin plants"', Browne, 'Botany for gentlemen', and Schiebinger, *Nature's Body*, pp. 11–39.

9. From *Præludia sponsaliorum plantarum*, quoted in Schiebinger, *Nature's Body*, p. 22.

10. Charles Alston and Samuel Goodenough quoted in Bewell, '"Jacobin plants"', p. 133.

11. Polwhele, *Unsex'd Females*, p. 8.

12. Quoted in Browne, 'Gentlemen of botany', p. 595 (from a letter). *The Loves of the Plants* was first published separately, and then as the second part of *The Botanic Garden*. The injunction to enlist imagination under the banner of science appeared in the double volume.

13. Darwin, *The Botanic Garden*, book 2 (*The Loves of the Plants*), pp. 4–5 (Canto I, ll. 51–6).

14. *Encyclopædia Britannica* quoted in Thornton, *The Temple of Flora*, p. 9.

15. Polwhele, *Unsex'd Females*, pp. 7–8.

16. Browne, 'Botany for gentlemen'; see also Danchin, 'Darwin's scientific and poetic purpose'.

17. Wakefield, 'Journals', 27 November 1798, 14 October 1799.

18. Wakefield, 'Journals', 14 October 1799. The main sources I have used for Wakefield's life are: Chapman, 'Wakefield'; Hill, 'Wakefield'; Shteir, 'Wakefield's natural history books' and *Cultivating Women*, pp. 81–9; Wakefield, *Mental Improvement*, pp. ix–xxi (Ann Shteir's introduction).

19. *Monthly Magazine*, November 1800, pp. 300–1.

20. Quoted in Hill, 'Wakefield', p. 4 (from the unpublished introduction to *Variety: Or Selections and Essays*).

21. Hawkins, *Letters on the Female Mind*, vol. 1, p. 12.

22. Benjamin, 'Elbow room', especially pp. 38–41.

23. Mullan, 'Gendered knowledge, gendered minds'; Myers, 'Science for women and children' and 'Fictionality'; Walters, 'Conversation pieces'.

24. Wakefield, 'Journals', 8 Apr 1799.

25. Wakefield, *Introduction to Botany*, p. 12.

26. Wakefield, *Mental Improvement*, p. 31.

27. Wakefield, *Introduction to Botany*, p. 17.

28. Wollstonecraft, *Vindication*, pp. 79, 154. Bewell, '"Jacobin plants"'.

29. Bermingham, *Learning to Draw*, pp. 202–27, George Brookshaw quoted on p. 206.

30. Ogilvie, 'Obligatory amateurs'.

11. Mary Shelley / Victor Frankenstein

1. The differences between the 1818 edition (the one I discuss) and the 1831 revised edition are best presented (by Marilyn Butler) in Shelley, *Frankenstein*. The major secondary sources to which I am indebted are Butler, 'Introduction' and Mellor, *Mary Shelley*; the best biography is Seymour, *Shelley*. In future footnotes I shall refer only to quotations from these.

2. *Edinburgh Magazine*, quoted in Baldick, *In Frankenstein's Shadow*, p. 57.

3. Baldick, *In Frankenstein's Shadow*.

4. St Clair, *Godwins and the Shelleys*, p. 436 (from *Pantheon*).

5. For detailed studies of less familiar novels, see Douthwaite, *Wild Girl*.

6. Letter from William Godwin to William Cole of 1802, quoted in Mellor, *Mary Shelley*, pp. 9–10. Seymour, *Shelley*, p. 54. My discussion of Marcet is based on Bahar, 'Jane Marcet' and Myers, 'Fictionality'.

7. Golinski, *Science as Public Culture*, pp. 188–235.

8. Coleridge quoted in Secord, *Victorian Sensation*, p. 46, n.11. Southey quoted in Myers, 'Fictionality', p. 48.

9. Armstrong, 'Jane Marcet', pp. 55–6. Farrell, 'Gentlewomen of science'. Neeley, *Mary Somerville*.

10. Radcliffe, *Female Advocate*, p. 399.

11. Secord, *Victorian Sensation*, pp. 42–6, and 'Scrapbook science'. Bermingham, *Learning to Draw*, pp. 145–64.

12. Hawkins, *Letters on the Female Mind*, vol. 1, p. 7. More, *Strictures on Education*, vol. 2, pp. 22–3. Benjamin, 'Elbow room'.

13. William Beckford, quoted in Baldick, *In Frankenstein's Shadow*, p. 56. Reviewer quoted in Hindle, 'Introduction', p. viii.

14. Jordanova, 'Melancholy reflection'.

15. John Barrow (1818), quoted in Butler, 'Introduction', pp. xxxiv–v.

16. Davy, *Works*, vol. 2, p. 319 (1802 lecture on chemistry), and vol. 8, p. 282 (1808 lecture on electrochemical science).

17. Davy, *Discourse*, p. 9.

18. Davy, *Discourse*, p. 9. John Robison (1803), quoted in Douthwaite, *Wild Girl*, p. 213.

19. Shelley, *Frankenstein*, pp. 38–9.

20. Shelley, *Frankenstein*, p. 30.

21. Shelley, *Frankenstein*, pp. 180, 16. Lawrence, 'Power and glory'.

22. Hindle, 'Introduction', p. xli.

23. Moers, 'Victorian Gothic'. Secord, 'Scrapbook science'. Bohls, *Women Travel Writers*, pp. 230–45.

24. Douthwaite, *Wild Girl*, especially pp. 211–22.

25. Barrell, *Birth of Pandora*, pp. 145–220, quotation on p. 148 from a standard translation of Hesiod.

26. Shelley, *Frankenstein*, p. 21.

27. In addition to previous sources, this discussion is based on Jordanova, 'Melancholy reflection', and Smith, 'Frankenstein and natural magic'.

28. Shelley, *Frankenstein*, p. 49. Cohen, *Fashioning Masculinity*.

29. Wollstonecraft, *Vindication*, p. 93.

Epilogue

1. Bal, 'Telling objects'.

2. Schiebinger, *Nature's Body*, pp. 40–74. Originally an Asian goddess, the multi-breasted figure was associated with Artemis by the Greeks, and with Diana by the Romans.

3. Boccaccio, *Famous Women*, 'Introduction' and pp. 14–17 (quotation on p. 17).

4. Anderson and Zinsser, *A History of Their Own*, vol. 1, pp. 54–6, 349–50; McLeod, *Order of the Rose*. The correct way to refer to her is Christine (not de Pizan), but since it is currently seen as patronising to refer to a woman only by her first name, I have repeated the full form.
5. Tomaselli, 'Collecting women'. For eighteenth-century England, see Guest, *Small Change*, pp. 49–50, 64–9, 168–72.

Bibliography

Abir-Am, Pnina G. and Outram, Dorinda (eds), *Uneasy Careers and Intimate Lives: Women in Science, 1789–1979* (New Brunswick and London: Rutgers University Press, 1987).

Adam, Charles, *Vie & œuvres de Descartes: étude historique* (Paris: le Cerf, 1910).

Aiton, E. J. *Leibniz: A Biography* (Bristol and Boston: Adam Hilger, 1985).

Åkerman, Susanna, *Queen Christina of Sweden and Her Circle: The Transformation of a Seventeenth-Century Philosophical Libertine* (Leiden: E. J. Brill, 1991).

Alic, Margaret, *Hypatia's Heritage: A history of Women in Science from Antiquity to the Late Nineteenth Century* (London: The Women's Press, 1986).

Anderson, Bonnie S. and Zinsser, Judith P., *A History of Their Own* (2 vols, New York and Oxford: Oxford University Press, 2000).

Armstrong, Eva V., 'Jane Marcet and her "Conversations on chemistry"', *Journal of Chemical Education* 15 (1938), 53–7.

Ashworth, William B. Jr, 'The Habsburg circle', in Bruce T. Moran (ed.), *Patronage and Institutions: Science, Technology and Medicine at the European Court, 1500–1750* (Woodbridge, Suffolk: Boydell Press, 1991), pp. 137–67.

Aubrey, John. *Brief Lives and Other Selected Writings*, ed. Anthony Powell (London: Cresset Press, 1949).

Backscheider, Paula R., *Reflections on Biography* (Oxford: Oxford University Press, 1999).

Bacon, Francis, *Works*, ed. J. Spedding, R. L. Ellis and D. D. Heath (7 vols, London: Longmans, London, 1870–6).

Badinter, Elisabeth, *Émilie, Émilie: l'ambition féminine au XVIIIème siècle* (Paris: Flammarion, 1983).

Bahar, Sara, 'Jane Marcet and the limits to public science', *British Journal for the History of Science* 34 (2001), 29–49.

Bal, Mieke, "Telling objects: a narrative perspective on collecting', in John Elsner and Roger Cardinal (eds), *Cultures of Collecting* (London: Reaktion, 1994), pp. 97–115.

Baldick, Chris, *In Frankenstein's Shadow: Myth, Monstrosity, and Nineteenth-Century Writing* (Oxford: Clarendon Press, 1987).

Barrell, John, *The Birth of Pandora and the Division of Knowledge* (London: Macmillan, 1992).

Barrett, Charlotte (ed.), *Diary and Letters of Madame d'Arblay* (4 vols, London: George Bell, 1891).

Barry, James, *Works* (2 vols, London: T. Cadell and W. Davies, 1809).

Bell, Whitfield J. Jr, '"All clear sunshine": new letters of Franklin and Mary Stevenson Hewson', *Proceedings of the American Philosophical Society* 100 (1956), 521–36.

Benjamin, Marina, 'Elbow room: women writers on science, 1790–1840', in Marina Benjamin (ed.), *Science and Sensibility: Gender and Scientific Enquiry, 1780–1945* (Oxford: Basil Blackwell, 1991), pp. 27–59.

Benjamin, Marina (ed.), *Science and Sensibility: Gender and Scientific Enquiry, 1780–1945* (Oxford: Basil Blackwell, 1991).

Benjamin, Marina (ed.), *A Question of Identity: Women, Science, and Literature* (New Brunswick: Rutgers University Press, 1993).

Bennett, J. A. '"On the power of penetrating into space": the telescopes of William Herschel', *Journal for the History of Astronomy* 7 (1976), 75–108.

Bennett, J. A. 'The Mechanics' Philosophy and the Mechanical Philosophy', *History of Science* 24 (1986), 1–28.

Bensaude-Vincent, Bernadette, 'Between history and memory: centennial and bicentennial images of Lavoisier', *Isis* 87 (1996), 48–199.

Beretta, Marco, 'At the source of western science: the organisation of experimentalism at the Accademia del Cimento (1657–1667)', *Notes and Records of the Royal Society* 54 (2000), 131–51.

Beretta, Marco, 'Chemical imagery and the Enlightenment of matter', in William R. Shea (ed.), *Science and the Visual Image in the Enlightenment* (Canton, MA: Science History Publications, 2000), pp. 57–88.

Beretta, Marco, *Imaging a Career in Science: The Iconography of Antoine Laurent Lavoisier* (Canton, MA: Science History Publications, 2001).

Bermingham, Ann, *Learning to Draw: Studies in the Cultural History of a Polite and Useful Art* (New Haven and London: Yale University Press, 2000).

Bertoloni Meli, D, 'Caroline, Leibniz, and Clarke', *Journal of the History of Ideas* 60 (1999), 469–86.

Besterman, Theodore, *Les lettres de la Marquise du Châtelet* (2 vols, Geneva: Institut et Musée Voltaire, 1958).

Bewell, Alan, '"Jacobin plants": botany as social theory in the 1790s', *Wordsworth Circle* 20 (1989), 132–9.

Biagioli, Mario, *Galileo Courtier: The Practice of Science in the Culture of Absolutism* (Chicago and London: University of Chicago Press, 1993).

Boccaccio, Giovanni, *Famous Women* (ed. and transl. Virginia Brown) (Cambridge, MA, and London: Harvard University Press, 2001).

Bohls, Elizabeth A., *Women Travel Writers and the Language of Aesthetics, 1716–1818* (Cambridge: Cambridge University Press, 1995).

Bordo, Susan (1987), 'The Cartesian masculinization of thought', in Sandra Harding and Jean F. O'Barr (eds), *Sex and Scientific Inquiry* (Chicago and London: University of Chicago Press, 1987), pp. 247–64.

Boyle, Robert, *The Works of the Honourable Robert Boyle,* ed. Thomas Birch (6 vols, London, 1772).

Brookner, Anita, *Jacques-Louis David* (London: Chatto & Windus, 1980).

Browne, Janet, 'Botany for gentlemen: Erasmus Darwin and *The loves of the plants*', *Isis* 80 (1989), 593–620.

Browne, Janet, *Charles Darwin: The Power of Place* (London: Jonathan Cape, 2002).

Brummer, Hans Henrik, 'Minerva of the north', in Marie-Louise Rodén (ed.), *Politics and Culture in the Age of Christina* (Stockholm: Suecoromana, 1997), pp. 77–92.

Brusati, Celeste, *Artifice and Illusion: The Art and Writing of Samuel van Hoogstraten* (Chicago and London: University of Chicago Press, 1995).

Bullard, Margaret, 'My small Newtonian sweeper – where is it now?', *Notes and Records of the Royal Society* 42 (1988), 139–48.

Butler, Marilyn, 'Introduction', in Mary Shelley, *Frankenstein or the Modern Prometheus: 1818 Text* (Oxford and New York: Oxford University Press, 1993), pp. ix–liii.

Cahn, Susan, *Industry of Devotion: The Transformation of Women's Work in England, 1500–1660* (New York: Columbia University Press, 1987).

Carroll, Bernice A. 'The politics of "originality": women and the class system of the intellect,' *Journal of Women's History* 2 (1990), 136–63.

Cavendish, Margaret, *Observations upon Experimental Philosophy* (London, 1668).

Centlivre, Susannah, *The Basset-Table: A Comedy* (London: for W. Mears, 1706).

Chapman, Georgiana, 'Priscilla Wakefield's life: draft chapters', Library of the London Society of Friends, Hazel Mews papers, Box 2, 21.

Clarke, Norma, *Dr Johnson's Women* (London and New York: Hambledon and London, 2000).

Clulee, Nicholas H., *John Dee's Natural Philosophy: Between Science and Religion* (London and New York: Routledge, 1988).

Cohen, Gustave, *Écrivains français en Hollande dans la première moitié du XVIIe siècle* (Paris: Édouard Champion, 1920).

Cohen, Michèle, *Fashioning Masculinity: National Identity and Language in the Eighteenth Century* (London and New York: Routledge, 1996).

Combremont, J. de Mestral, *Albertine Necker de Saussure, 1766–1841* (Lausanne: Librairie Payot, 1946).

Connor, Elizabeth, 'Mme. Lepaute, an eighteenth-century computer', *Astronomical Society of the Pacific*, Leaflet 189 (November 1944).

Conway, Anne, *The Principles of the Most Ancient and Modern Philosophy*, ed. Allison P. Coudert and Taylor Corse (Cambridge: Cambridge University Press, 1996).

Cook, Alan, *Edmund Halley: Charting the Heavens and the Seas* (Oxford: Clarendon Press, 1998).

Cook, Alan, 'Johann and Elizabeth Hevelius, astronomers of Danizig', *Endeavour* 24 (2000), 8–12.

Cooter, Roger and Pumfrey, Stephen, 'Separate spheres and public places: reflections on the history of science popularization and science in popular culture', *History of Science* 32 (1994), 237–67.

Cormack, Lesley B., 'Twisting the lion's tail: practice and theory at the court of Henry Prince of Wales', in Bruce T. Moran (ed.), *Patronage and Institutions: Science, technology and Medicine at the European Court, 1500–1750* (Woodbridge, Suffolk: Boydell Press, 1991), pp. 67–83.

Costa, Shelley, 'The *Ladies' Diary*: Gender, mathematics and civil society in early eighteenth-century England', *Osiris* 17 (2002), 49–73.

Coudert, Allison P., *Leibniz and the Kabbalah* (Dordrecht, Boston and London: Kluwer, 1995).

Creese, Mary R. S., *Ladies in the Laboratory: American and British Women in Science, 1800–1900: A Survey of Their Contributions to Research* (Lanham, MD, and London: Scarecrow Press, 1998).

Cunningham, Andrew and Williams, Perry, 'De-centring the "big picture": *The Origins of Modern Science* and the modern origins of science', *British Journal for the History of Science* 26 (1993), 407–32.

Cust, Lionel and Malloch, Archibald, 'Portraits by Carlo Dulci and S. van Hoogstraten', *Burlington Magazine* 29 (1916), 292–3.

Dalton, R. and Hamer, S. H. *The Provincial Token-Coinage of the 18ᵗʰ Century* (2 vols, London, 1910–14).

Danchin, Pierre, 'Erasmus Darwin's scientific and poetic purpose in *The Botanic Garden*', in Siergo Rossi (ed.), *Science and Imagination in XVIIIth-Century British Culture* (Milan: Edizioni Unicopoli, 1987), pp. 133–50.

Darwin, Charles, *The Correspondence of Charles Darwin: Volume 4, 1847–1850*, ed. Frederick Burkhardt and Sydney Smith (Cambridge: Cambridge University Press, 1988).

Darwin, Erasmus, *The Botanic Garden* (Yorkshire: Scolar Press, 1973) (facsimile of 1791 edition).

Davy, Humphry, *A Discourse, Introductory to a Course of Lectures on Chemistry, Delivered in the Theatre of the Royal Institution on the 21ˢᵗ of January, 1802* (London: J. Johnson, 1802).

Davy, Humphry, *The Collected Works of Sir Humphry Davy*, ed. John Davy (9 vols, London: Smith, Elder, 1839–40).

De Nanteuil, Luc., *Jacques-Louis David* (London: Thames and Hudson, 1990).

Dee, John, 'Preface', in *The Elements of Geometrie of the Most Auncient Philosopher Euclide of Megara*, transl. Henry Billingsley (London, 1570).

Descartes, Catherine, "Relation de la mort de M Descartes, le philosophe," in Dominique Bouhours, *Recueil de vers chorisis* (Paris, 1693), pp. 129–39.

Descartes, René, *Lettres sur la morale: correspondence avec la princesse Élisabeth, Chanut et la reine Christine*, ed. Jacques Chevalier (Paris: Boivin, 1931).

Descartes, René, *Principles of Philosophy*, transl. Valentine Rodger Miller and Reese P. Miller (Dordrecht: D. Reidel, 1983).

Donovan, Arthur, *Antoine Lavoisier: Science, Administration, and Revolution* (Oxford and Cambridge, MA: Blackwell, 1993).

Douthwaite, Julia V., *The Wild Girl Natural Man and the Monster: Dangerous Experiments in the Age of Enlightenment* (Chicago and London: University of Chicago Press, 2002).

Du Châtelet, Émilie, *Institutions de physique* (Paris: Prault, 1740).

Du Châtelet, Émilie, *Principes mathématiques de la philosophie naturelle* (2 vols, Paris: Éditions Jacques Gabay, 1990 facsimile of 1759 edition).

Duportal, Jeanne, *Bernard Picart 1673 à 1733* (Paris and Brussels: Éditions g. Van Oest, 1928).

Duran, Jane, 'Anne Viscountess Conway: a seventeenth century rationalist', *Hypatia* 4/1 (1989), 64–79.

Duveen, Debis I., 'Madame Lavoisier, 1758–1836', *Chymia* 4 (1953), 13–29.

Easlea, Brian, *Witch-Hunting, Magic and the New Philosophy: An Introduction to the Debates of the Scientific Revolution, 1450–1750* (Brighton: Harvester Press, 1980).

Easlea, Brian, *Fathering the Unthinkable: Masculinity, Scientists and the Nuclear Arms Race* (London: Pluto Press, 1983).

Edwards, Samuel, *The Divine Mistress* (London: Cassell, 1971).

Ehrman, Esther, *Mme du Châtelet: Scientist, Philosopher and Feminist of the Enlightenment* (Leamington Spa: Berg, 1986).

Evans, James, 'Fraud and illusion in the anti-Newtonian rear guard', *Isis* 87 (1996), 74–107.

Fara, Patricia, *Newton: The Making of Genius* (London: Macmillan, 2002).

Fara, Patricia, 'Elizabeth Tollet: a new Newtonian woman', *History of Science* 40 (2002), 169–87.

Farrell, Honor Cecilia, "Gentlewomen of science: the role of women in the London scientific elite, 1800–1875', M.Phil. dissertation, University of Leeds, 1994.

Farrington, Benjamin, *The Philosophy of Francis Bacon: An Essay on its Development from 1603 to 1609 with New Translations of Fundamental Texts* (Liverpool: Liverpool University Press, 1964).

Faujas de Saint Fond, Barthelemy, *A Journey through England and Scotland to the Hebrides in 1784*, ed. Archibald Geikie (2 vols, Glasgow: Hugh Hopkins, 1907).

Fenton, Edward (ed.). *The Diaries of John Dee* (Charlbury: Day Books, 1998).

Ferguson, James, *The Young Gentleman and Lady's Astronomy, Familiarly Explained in Ten Dialogues between Neander and Eudosia* (Dublin, 1768).

French, Peter J., *John Dee: The World of an Elizabethan Magus* (London: Routledge & Kegan Paul, 1972).

Findlen, Paula, 'A forgotten Newtonian: women and science in the Italian provinces', in William Clark, Jan Golinski and Simon Schaffer (eds), *The Sciences in Enlightened Europe* (Chicago and London: University of Chicago Press, 1999), pp. 313–49.

Frankel, Lois, 'Damaris Cudworth Masham: a seventeenth century feminist philosopher', *Hypatia* 4/1 (1989), 80–90.

Gates, Barbara T. and Shteir, Ann B. (eds), *Natural Eloquence: Women Reinscribe Science* (Wisconsin: University of Wisconsin Press, 1997).

Gaukroger, Stephen, *Descartes: An Intellectual Biography* (Oxford: Clarendon Press, 1995).

Gerhardt, C. I. (ed.). *Die philosophischen Schriften von G. W. Leibniz* (7 vols, Berlin, 1875–90).

Glanvill, Joseph, *The Vanity of Dogmatizing* (London, 1661).

Godfrey, Elizabeth, *A Sister of Prince Rupert* (London and New York: John Lane, 1909).

Golinski, Jan, *Science as Public Culture: Chemistry and Enlightenment in Britain, 1760–1820* (Cambridge: Cambridge University Press, 1992).

Golinski, Jan, 'The care of the self and the masculine birth of science', *History of Science* 40 (2002), 125–45.

Goodman, Dena, *The Republic of Letters: A Cultural History of the French Enlightenment* (Ithaca and London: Cornell University Press, 1994).

Gould, Stephen Jay, 'The invisible woman', in Barbara T. Gates and Ann B. Shteir (eds), *Natural Eloquence: Women Reinscribe Science* (Wisconsin: University of Wisconsin Press, 1997), pp. 27–39.

Grant, Douglas, *Margaret the First: A Biography of Margaret Cavendish, Duchess of Newcastle, 1623–1673* (London: Rupert Hart-Davis, 1957).

Grimaux, Édouard, *Lavoisier: 1743–1794* (Paris: Germer Baillière, 1888).

Guest, Harriet, *Small Change: Women, Learning, Patriotism, 1750–1810* (Chicago and London: University of Chicago Press, 2000).

Guizot, François, 'Madame de Rumford (1758–1836)', in Guizot, *Mélanges biographiques et littéraires* (Paris: Michel Lévy, 1868), pp. 49–88.

Hannay, Margaret P., "How I these studies prize": the Countess of Pembroke and Elizabethan science', in Lynette Hunter and Sarah Hutton (eds), *Women, Science and Medicine 1500–1700: Mothers and Sisters of the Royal Society* (Stroud: Sutton Publishing, 1997), pp. 108–21.

Harkness, Deborah E., 'Managing an experimental household: the Dees of Mortlake and the practice of natural philosophy', *Isis* 88 (1997), 247–62.

Harkness, Deborah E., *John Dee's Conversations with Angels: Cabala, Alchemy, and the End of Nature* (Cambridge: Cambridge University Press, 1999).

Harris, Frances, "Living in the neighbourhood of science: Mary Evelyn, Margaret Cavendish and the Greshamites', in Lynette Hunter and Sarah Hutton (eds), *Women, Science and Medicine 1500–1700: Mothers and Sisters of the Royal Society* (Stroud: Sutton Publishing, 1997), pp. 198–217.

Harth, Erica, *Cartesian Women: Versions and Subversions of Rational Discourse in the Old Regime* (Ithaca and London: Cornell University Press, 1992).

Hartman, Mary S., *Victorian Murderesses: A True History of Thirteen Respectable French and English Women Accused of Unspeakable Crimes* (London: Robson Books, 1977).

Hawkins, Laetitia, *Letters on the Female Mind, its Powers and Pursuits* (2 vols, London: Hookham and Carpenter, 1793).

Heilbron, John L., 'Franklin as an enlightened natural philosopher', in J. A. Leo Lemay (ed.), *Reappraising Benjamin Franklin: A Bicentennial Perspective* (Newark: University of Delaware Press, 1993), pp. 196–220.

Heilbrun, Carolyn G. *Writing a Woman's Life* (London: The Women's Press, 1988).

Henry, Madeleine M., *Prisoner of History: Aspasia of Miletus and her Biographical Tradition* (New York and Oxford: Oxford University Press, 1995).

Herschel, Caroline, 'An account of the discovery of a comet', *Philosophical Transactions* 84 (1794), 1.

Herschel, Mary Cornwallis, *Memoir and Correspondence of Caroline Herschel* (London: John Murray, 1876).

Hevelius, Johannes, *Prodromus astronomiæ firmamentum Sobiescianum sive uranographia* (Danzig, 1690).

Hill, Bridget, 'Priscilla Wakefield as a writer of children's educational books', *Women's Writing* 4 (1997), 3–15.

Hindle, Maurice, 'Introduction', in Mary Shelley, *Frankenstein or the Modern Prometheus* (London: Penguin, 1992), pp. vii–xliii.

Hoffman, Roald, 'Mme Lavoisier', *American Scientist* 90 (January–February 2002), 22–4.

Hoskins, Michael, 'Caroline Herschel's comet sweepers', *Journal for the History of Astronomy* 12 (1981), 27–34.

Hunter, Lynette, 'Sisters of the Royal Society: the circle of Katherine Jones, Lady Ranelagh', in Lynette Hunter and Sarah Hutton (eds), *Women, Science and Medicine 1500–1700: Mothers and Sisters of the Royal Society* (Stroud: Sutton Publishing, 1997), pp. 178–97.

Hunter, Lynette and Hutton, Sarah, *Women, Science and Medicine 1500–1700: Mothers and Sisters of the Royal Society* (Stroud: Sutton Publishing, 1997).

Hunter, Michael, *Science and Society in Restoration England* (Cambridge: Cambridge University Press, 1981).

Hutton, Sarah, 'Of physic and philosophy: Helmontian medicine in Restoration England', in Ole Peter Grell and Andrew Cunningham (eds), *Religio medici: Medicine and Religion in Seventeenth-Century England* (Aldershot: Scolar Press, 1996), pp. 228–46.

Hutton, Sarah, 'The riddle of the sphinx: Francis Bacon and the emblems of science', in Lynette Hunter and Sarah Hutton (eds), *Women, Science and Medicine 1500–1700: Mothers and Sisters of the Royal Society* (Stroud: Sutton Publishing, 1997), pp. 7–28.

Hutton, Sarah, 'Anne Conway, Margaret Cavendish and seventeenth-century scientific thought', in Lynette Hunter and Sarah Hutton (eds), *Women, Science and Medicine 1500–1700: Mothers and Sisters of the Royal Society* (Stroud: Sutton Publishing, 1997), pp. 218–34.

Iliffe, Rob, '"Aplatisseur du monde et de Cassini": Maupertuis, precision measurement, and the shape of the earth in the 1730s', *History of Science* 31 (1993), 335–75.

Iliffe, Rob, 'The masculine birth of time: temporal frameworks of early modern natural philosophy', *British Journal for the History of Science* 33 (2000), 427–53.

Iliffe, Rob and Willmoth, Frances, 'Astronomy and the domestic sphere: Margaret Flamsteed and Caroline Herschel as assistant-astronomers', in Lynette Hunter and Sarah Hutton (eds), *Women, Science and Medicine 1500–1700: Mothers and Sisters of the Royal Society* (Stroud: Sutton Publishing, 1997), pp. 235–65.

Iltis (Merchant), Carolyn, 'Madame du Châtelet's metaphysics and mechanics', *Studies in History and Philosophy of Science* 8 (1977), 29–48.

Israel, Jonathan I., *Radical Enlightenment: Philosophy and the Making of Modernity 1650–1750* (Oxford: Oxford University Press, 2001).

Janik, Linda Gardiner, 'Searching for the metaphysics of science: the structure and composition of Mme du Châtelet's *Institutions de physique*, 1737–1740', *Studies on Voltaire and the Eighteenth Century* 201 (1982), 85–113.

Jones, Kathleen, *A Glorious Fame: The Life of Margaret Cavendish, Duchess of Newcastle, 1623–1673* (London: Bloomsbury, 1988).

Jordanova, Ludmilla, *Sexual Visions: Images of Gender in Science and Medicine between the Eighteenth and Twentieth Centuries* (Madison: University of Wisconsin Press, 1989).

Jordanova, Ludmilla, 'Gender and the historiography of science', *British Journal for the History of Science* 26 (1993), 469–83.

Jordanova, Ludmilla, 'Melancholy reflection: constructing an identity for unveilers of nature', in Stephen Bann (ed.), *Frankenstein Creation and Monstrosity* (London: Reaktion Books, 1994), pp. 60–76.

Jordanova, Ludmilla, *Nature Displayed: Gender, Science and Medicine 1760–1820* (London and New York: Longman, 1999).

Kant, Immanuel, 'Observations on the feeling for the beautiful and the sublime', in Gabrielle Rabel, *Kant* (Oxford: Clarendon Press, 1963), pp. 61–3.

Kargon, Robert Hugh, *Atomism in England from Hariot to Newton* (Oxford: Clarendon Press, 1966).

Kass-Simon, G. and Farnes, Patricia (eds), *Women of Science: Righting the Record* (Bloomington: Indiana University Press, 1990).

Keller, Evelyn Fox, *Reflections on Gender and Science* (Yale: Yale University Press, 1985).

Keller, Evelyn Fox, 'Secrets of God, nature, and life', in her *Secrets of Life, Secrets of Death: Essays on Gender, Language, and Science* (New York and London: Routledge, 1992), pp. 56–72 (originally published as *History of the Human Sciences* 3 (1990), 229–42).

Klopp, Onno, *Die Werke von Leibniz* (11 vols, Hanover: Klindworth, 1864–84).

Koerner, Lisbet, 'Women and utility in Enlightenment science', *Configurations* 3 (1995), 233–55.

Koerner, Lisbet, *Linnaeus: Nature and Nation* (Cambridge, MA, and London: Harvard University Press, 1999).

Kohlstedt, Sally Gregory and Longino, Helen E., *Women, Gender, and Science: New Directions, Osiris* 12 (1997).

Lalande, Joseph de, *Ladies' Astronomy*, transl. W. Pengree (London, 1815).

Laurence, Anne, *Women in England 1500–1760: A Social History* (London: Weidenfeld & Nicolson, 1994).

Lawrence, Christopher, 'The power and the glory: Humphry Davy and Romanticism', in Andrew Cunningham and Nicholas Jardine (eds), *Romanticism and the Sciences* (Cambridge: Cambridge University Press, 1990), pp. 213–27.

Lubbock, Constance A., *The Herschel Chronicle: the Life-Story of William Herschel and his Sister Caroline Herschel* (Cambridge: Cambridge University Press, 1933).

McKie, Douglas, *Antoine Lavoisier: Scientist, Economist, Social Reformer* (London: Constable, 1952).

McLeod, Enid, *The Order of the Rose: The Life and Ideas of Christine de Pizan* (London: Chatto & Windus, 1976).

MacPike, Eugene Fairfield, *Hevelius, Flamsteed and Halley: Three Contemporary Astronomers and their Mutual Relations* (London: Taylor & Francis, 1937).

Makin, Bathsua, *An Essay to Revive the Ancient Education of Gentlewomen, in Religion, Manners, Arts & Tongues* (London, 1673).

Martensen, Robert, 'The transformation of Eve: women's bodies, medicine and culture in early modern England', in Roy Porter and Mikuláš Teich (eds), *Sexual Knowledge, Sexual Science: The History of Attitudes to Sexuality* (Cambridge: Cambridge University Press, 1994).

Martin, Benjamin, *The Young Gentlemen's and Ladies Philosophy* (2 vols, London, 1759–63).

Mattern, Ruth, 'Descartes's correspondence with Elizabeth: concerning both the union and distinction of mind and body', in Michael Hooker (ed.), *Descartes: Critical and Interpretive Essays* (Baltimore and London: Johns Hopkins University Press, 1978), pp. 212–22.

Mazzotti, Massimo, 'Maria Gaetana Agnesi: mathematics and the making of the Catholic Enlightenment', *Isis* 92 (2001), 657–83.

Mellor, Anne K., *Mary Shelley: Her Life her Fiction her Monsters* (New York and London: Routledge, 1988).

Merchant, Carolyn, 'The vitalism of Anne Conway: its impact on Leibniz's concept of the monad', *Journal of the History of Philosophy* 17 (1979), 255–69.

Merchant, Carolyn, *The Death of Nature: Women, Ecology, and the Scientific Revolution* (London: Wildwood House, 1980).

Merchant, Carolyn, 'Isis' consciousness raised', *Isis* 73 (1982), 398–409.

Merton, Robert, *On the Shoulders of Giants: A Shandean Postscript* (Chicago and London: University of Chicago Press, 1993).

Millburn, John R., *Adams of Fleet Street, Instrument Makers to King George III* (Aldershot: Ashgate, 2000).

Miller, James, *The Humours of Oxford* (Dublin, 1730).

Mintz, Samuel I., 'The Duchess of Newcastle's visit to the Royal Society', *Journal of English and Germanic Philology* 51 (1952), 168–76.

Moers, Ellen, 'Female Gothic', in George Levine and U. C. Knoepflmacher (eds), *The Endurance of Frankenstein: Essays on Mary Shelley's Novel* (Berkeley, Los Angeles and London: University of California Press, 1974), pp. 77–87.

Montagu, Mary Wortley, *The Complete Letters of Lady Mary Wortley Montagu* (3 vols, Oxford: Oxford University Press, 1965–7).

More, Hannah, *Strictures on the Modern System of Female Education* (2 vols, London, 1799).

More, Henry, *An Antidote against Atheisme* (London, 1653).

Mozans, H. J., *Woman in Science* (New York and London: D. Appleton, 1913).

Mullan, John, 'Gendered knowledge, gendered minds: women and Newtonianism, 1690–1760', in Marina Benjamin (ed.), *A Question of Identity: Women, Science and Literature* (New Jersey: Rutgers University Press, 1993), pp. 41–56.

Myers, Greg, "Science for women and children: the dialogue of popular science in the nineteenth century', in John Christie and Sally Shuttleworth (eds), *Nature Transfigured: Science and Literature, 1700–1900* (Manchester and New York: Manchester University Press, 1989), pp. 171–200.

Myers, Greg, 'Fictionality, demonstration, and a forum for popular science: Jane Marcet's *Conversations on chemistry*', in Barbara T. Gates and Ann B. Shteir (eds), *Natural Eloquence: Women Reinscribe Science* (Wisconsin: University of Wisconsin Press, 1997), pp. 43–60.

Néel, Marguerite, *Descartes et la princesse Elisabeth* (Paris: Elzévir, 1946).

Neeley, Kathryn A., *Mary Somerville: Science, Illumination, and the Female Mind* (Cambridge: Cambridge University Press, 2001).

Nicolson, Marjorie Hope and Hutton, Sarah (eds), *The Conway Letters: The Correspondence of Anne, Viscountess Conway, Henry More, and their Friends 1642–1684* (Oxford: Clarendon Press, 1992).

Nye, Robert A., 'Medicine and science as masculine "fields of honor"', *Osiris* 12 (1997), 60–79.

Ogilvie, Marilyn Bailey, 'Caroline Herschel's contributions to astronomy', *Annals of Science* 32 (1975), 149–61.

Ogilvie, Marilyn Bailey. *Women in Science: Antiquity through the Nineteenth Century: A Biographical Dictionary with Annotated Bibliography* (Cambridge, MA: MIT Press, 1996).

Ogilvie, Marilyn Bailey, 'Obligatory amateurs: Annie Maunder (1869–1947) and British women astronomers at the dawn of professional astronomy', *British Journal for the History of Science* 33 (2000), 67–84.

Ogilvie, Marilyn and Harvey, Joy (eds), *The Biographical Dictionary of Women in Science: Pioneering Lives from Ancient Times to the Mid-20th Century* (2 vols, New York and London: Routledge, 2000).

Oldenburg, Henry, *The Correspondence of Henry Oldenburg*, transl. and ed. A. Rupert Hall and Marie Boas Hall (9 vols, Madison: University of Wisconsin Press, 1966–73).

Osen, Lynn M., *Women in Mathematics* (Cambridge, MA, and London: MIT Press, 1974).

Outram, Dorinda, 'Before objectivity: wives, patronage, and cultural reproduction in early nineteenth-century French science', in Pnina G. Abir-Am and Dorinda Outram (eds), *Uneasy Careers and Intimate Lives: Women in Science, 1789–1979* (New Brunswick and London: Rutgers University Press, 1987), pp. 19–30.

Poirier, Jean-Pierre, *Lavoisier: Chemist, Biologist, Economist* (Philadelphia: University of Pennsylvania Press, 1993).

Poirier, Jean-Pierre, *Histoire des femmes en science en France du Moyen Age à la Révolution* (Paris: Pygmalion, 2002).

Polwhele, Richard, *The Unsex'd Females: A Poem* (London, 1798).

Popkin, Richard H., 'The spiritualistic cosmologies of Henry More and Anne

Conway', in Sarah Hutton (ed.), *Henry More (1614–1687): Tercentenary Studies* (Dordrecht, Boston and London: Kluwer, 1990), pp. 97–114.

Porter, Roy, 'A touch of danger: the man-midwife as sexual predator', in George Rousseau and Roy Porter (eds), *Sexual Underworld of the Enlightenment* (Manchester: Manchester University Press, 1987), pp. 206–32.

Pumfrey, Stephen, 'Who did the work? Experimental philosophers and public demonstrators in Augustan England', *British Journal for the History of Science* 28 (1995), 131–56.

Pycior, Helena M., 'Pierre Curie and "his eminent collaborator M^{me}Curie": complementary partners', in Helena M. Pycior, Nancy G. Slack and Pnina G. Abir-Am (eds), *Creative Couples in the Sciences* (New Brunswick: Rutgers University Press, 1997) pp. 39–56.

Pycior, Helena M., Slack, Nancy G. and Abir-Am, Pnina G. (eds), *Creative Couples in the Sciences* (New Brunswick: Rutgers University Press, 1997).

Radcliffe, Mary Ann, *The Female Advocate: Or, An Attempt to Recover the Rights of Women from Male Usurpation* (London: 1799).

Rang, Brita, '"An exceptional mind": the learned Anna Maria van Schurman', in Mirjam de Baar *et al.* (eds), *Choosing the Better Part: Anna Maria van Schurman (1607–1678)* (Dordrecht, Boston and London: Kluwer, 1996), pp. 23–42.

Rayner-Canham, G. W. and Frenette, H. 'Some French women chemists', *Education in Chemistry* 22 (1985), 176–8.

Reay, Barry, *Watching Hannah: Sexuality, Horror and Bodily De-formation in Victorian England* (London: Reaktion Books, 2002).

Rebière, A., *Les femmes dans la science* (Paris: Nony, 1894).

Reynolds, Myra, *The Learned Lady in England 1650–1760* (Gloucester, MA: Peter Smith, 1964).

Richards, Joan, *Angles of Reflection: A Memoir of Logic and a Mother's Love* (New York: W. H. Freeman, 2000).

Richardson, George, *Iconology: London 1779* (2 vols, New York and London: Garland Publishing Inc. 1979) (facsimile reprint of 1779 edition).

Richmond, Marsha L., 'Women in the early history of genetics: William Bateson and the Newnham College Mendelians, 1900–1910', *Isis* 92 (2001), 55–90.

Rogers, Katharine M., *The Troublesome Helpmate: A History of Misogyny in Literature* (Seattle and London: University of Washington Press, 1966).

Rossiter, Margaret, *Women Scientists in America: Struggles and Strategies to 1948* (Baltimore: Johns Hopkins University Press, 1982).

St Clair, William, *The Godwins and the Shelleys: The Biography of a Family* (London and Boston: Faber & Faber, 1989).

Saxl, Fritz, 'Veritas filia temporise', in Raymond Klibansky and H. J. Paton (eds), *Philosophy and History: Essays Presented to Ernst Cassirer* (Oxford: Clarendon Press, 1936), pp. 197–222.

Schaffer, Simon, 'Herschel in Bedlam: natural history and stellar astronomy', *British Journal for the History of Science* 13 (1980), 211–39.

Schaffer, Simon, 'Halley, Delisle, and the making of the comet', in Norman J. W. Thrower (ed.), *Standing on the Shoulders of Giants: A Longer View of Newton and Halley* (Berkeley, Los Angeles and Oxford: University of California Press, 1990), pp. 254–98.

Scheler, Lucien, 'Deux lettres inédites de Mme Lavoisier', *Revue d'Histoire des Sciences* 38 (1985), 121–30.

Schiebinger, Londa, 'Maria Winkelmann at the Berlin Academy: a turning point for women in science', *Isis* 78 (1987), 174–200.

Schiebinger, Londa, *The Mind Has No Sex? Women in the Origins of Modern Science* (Cambridge and London: Harvard University Press, 1989).

Schiebinger, Londa, *Nature's Body: Gender in the Making of Modern Science* (Boston: Beacon Press, 1993).

Schiebinger, Londa, *Has Feminism Changed Science?* (Cambridge, MA, and London: Harvard University Press, 1999).

Schnapper, Antoine, *David* (New York: Alpine Fine Arts, 1980).

Secord, Anne, 'Science in the pub: artisan botanists in early nineteenth-century Lancashire', *History of Science* 32 (1994), 269–315.

Secord, James A., *Victorian Sensation: The Extraordinary Publication, Reception, and Secret Authorship of* Vestiges of the Natural History of Creation (Chicago and London: University of Chicago Press, 2000).

Secord, James A., 'Scrapbook science: composite caricatures in late Georgian England', forthcoming.

Seymour, Miranda, *Mary Shelley* (London: Picador, 2001).

Shapin, Steven, '"The house of experiment in seventeenth-century England"', *Isis* 79 (1988), 373–404.

Shapin, Steven, 'Descartes the doctor: rationalism and its therapies', *British Journal for the History of Science* 33 (2000), 131–54.

Shapiro, Lisa, 'Princess Elizabeth and Descartes: the union of soul and body in the practice of philosophy', *British Journal for the History of Philosophy* 7 (1999), 503–20.

Shelley, Mary, *Frankenstein or the Modern Prometheus: 1818 Text* (Oxford and New York: Oxford University Press, 1993).

Shteir, Ann B., 'Priscilla Wakefield's natural history books', in Alwyne Wheeler and James H. Price (eds), *From Linnaeus to Darwin: Commentaries on the History of Biology and Geology* (London: Society for the History of Natural History, 1985).

Shteir, Ann B., *Cultivating Women, Cultivating Science: Flora's Daughters and Botany in England 1760 to 1860* (Baltimore and London: Johns Hopkins University Press, 1996).

Siegfried, Susan L., 'Boilly and the frame-up of *trompe l'œil*,' *Oxford Art Journal* 15/2 (1992), 27–37.

Simon, Jonathan, 'Mineralogy and mineral collections in 18th-century France', *Endeavour* 26 (2002), 132–6.

Smeaton, William, 'Monsieur and Madame Lavoisier in 1789: the chemical revolution and the French Revolution', *Ambix* 36 (1989), 1–4.

Smeaton, William, 'Madame Lavoisier, P. S. and E. I. du Pont de Nemours and the publication of Lavoisier's "Mémoires de Chimie"', *Ambix* 36 (1989), 22–30.

Smith, Bonnie, *The Gender of History: Men, Women, and Historical Practice* (Cambridge, MA, and London: Harvard University Press, 1998).

Smith, Crosbie, 'Frankenstein and natural magic', in Stephen Bann (ed.), *Frankenstein Creation and Monstrosity* (London: Reaktion Books, 1994), pp. 39–59.

Sophia, Electress of Hanover, *Memoirs*, transl. H. Forester (London: Richard Bentley, 1888).

Sorbière, Samuel, *Sorberiana, ou bons mots, rencontres agreables, pensées judicieuses, et observations curieuses* (Paris, 1694).

Sørenson, Madeleine Pinault, 'La méditation dans les ruines', in *Charles-Louis Clérisseau (1721–1820): dessins du muse de l'Ermitage, Saint-Pétersbourg* (Paris: Réunion des Musées Nationaux, 1995), pp. 51–4.

South, James, 'An address delivered at the Annual General Meeting of the Astronomical Society of London, on February 8, 1828, on presenting the Honorary medal to Miss Caroline Herschel', *Memoirs of the Royal Astronomical Society* 3 (1829), 409–12.

Sprat, Thomas, *The History of the Royal-Society of London, for the Improving of Natural Knowledge* (London, 1667).

Tabor, Margaret E., *Pioneer Women* (4 vols, London: Sheldon Press, 1925–33).

Terrall, Mary, 'Émilie du Châtelet and the gendering of science', *History of Science* 33 (1995), 283–310.

Thornton, Robert, *The Temple of Flora*, ed. Ronald King (London: Weidenfeld & Nicolson, 1981).

Tollet, Elizabeth, *Poems on Several Occasions* (London: for T. Lownds, *c.* 1760).

Tomaselli, Sylvana, 'Collecting women: the female in scientific biography', *Science as Culture* 4 (1988), 95–106.

Turnbull, H. W. *et al.*, *The Correspondence of Isaac Newton* (7 vols, Cambridge: Cambridge University Press, 1959–77).

Turner, A. J., *Science and Music in Eighteenth Century Bath* (Bath: University of Bath, 1977).

Uglow, Jenny, *The Lunar Men: The Friends who Made the Future, 1730–1810* (London: Faber & Faber, 2002).

Veldman, Ilja M. and Luijten, Ger, *The New Hollstein Dutch & Flemish Etchings, Engravings and Woodcuts 1450–1700: Marten van Heemskerck* (2 vols, Roosendaal: Koninklijke Van Poll, 1993–4).

Vickery, Amanda, 'Golden age to separate spheres: a review of the categories and chronology of English women's history', *Historical Journal* 36/2 (1993), 383–414.

Vickery, Amanda, *The Gentleman's Daughter: Women's Lives in Georgian England* (New Haven and London: Yale University Press, 1998).

Vidal, Mary, 'David among the moderns: art, science, and the Lavoisiers', *Journal of the History of Ideas* 56 (1995), 595–623.

Vignoles, Alphonse, 'Eloge de Madame Kirch à l'occasion de laquelle on parle de quelques autres Femmes & d'un Paisan Astronomes', *Bibliothèque Germanique* 3 (1721), 155–83.

Voltaire, *Élémens de la philosophie de Newton, mis à la portée de tout le monde* (Amsterdam: Etienne Ledet, 1738).

Voltaire, *Correspondence*, ed. Theodore Besterman (107 vols, Geneva: Institut et Musée Voltaire, 1953–65).

Voltaire, *Letters on England*, trans. L. Tancock (Harmondsworth: Penguin 1980).

Vrooman, Jack Rochford, *René Descartes: A Biography* (New York: G. P. Putnam's Sons, 1970).

Wade, Ira O., *Voltaire and Madame du Châtelet: An Essay on the Intellectual Activity at Cirey* (Princeton: Princeton University Press, 1941).

Wade, Ira O., *The Intellectual Development of Voltaire* (Princeton: Princeton University Press, 1969).

Wagner-Martin, Linda, *Telling Women's Lives: The New Biography* (New Brunswick: Rutgers University Press, 1994).

Wakefield, Priscilla, 'Journals 1796–1816'. Library of the London Society of Friends, Hazel Mews papers, Box 1, 1.

Wakefield, Priscilla, *An Introduction to Botany, in a Series of Familiar Letters* (London, 1796).

Wakefield, Priscilla, *Mental Improvement*, ed. Ann B. Shteir (East Lansing: Colleagues Press, 1995).

Wallington, Emma, 'The physical and intellectual capacities of woman equal to those of man', *Anthropologia* 1 (1874), 552–65.

Walters, Alice N. 'Conversation pieces: science and politeness in eighteenth-century England', *History of Science* 35 (1997), 121–54.

Walters, Robert L., 'The allegorical engravings in the Ledet-Lesbordes edition of the *Élemens de la philosophie de Newton*', in R. J. Howells, A. Mason, H. T. Mason and D. Williams (eds), *Voltaire and His World: Studies Presented to W. H. Barber* (Oxford: Voltaire Foundation, 1985), pp. 27–49.

Weisbard, P. H. (ed.), *The History of Women and Science, Health, and Technology* (Madison: University of Wisconsin Press, 1993).

Westfall, Richard S., *Never at Rest: A Biography of Isaac Newton* (Cambridge: Cambridge University Press, 1980).

Wilson, Adrian, *The Making of Man-midwifery: Childbirth in England, 1660–1770* (Cambridge, MA: Harvard University Press, 1995).

Winkler, Mary G. and van Helden, Albert. 'Johannes Hevelius and the visual language of astronomy', in J. V. Field and Frank A. J. L. James (eds), *Renaissance and Revolution: Humanists, Scholars, Craftsmen and Natural Philosophers in Early Modern Europe* (Cambridge: Cambridge University Press, 1993), pp. 97–116.

Wollstonecraft, Mary, *Vindication of the Rights of Woman* (Harmondsworth: Penguin, 1975).

Woolf, Virginia, *A Room of One's Own* (Harmondsworth: Penguin, 1972).

Young, Arthur, *Travels in France during the Years 1787, 1788 and 1789*, ed. Jeffry Kaplow (New York: Anchor Books, 1969).

Zedler, Beatrice H., 'The three princesses', *Hypatia* 4/1 (1989), 28–63.

Zinsser, Judith P., 'Émilie du Châtelet: Genius, gender, and intellectual authority', in Hilda L. Smith (ed.), *Women Writers and the Early Modern British Political Tradition* (Cambridge: Cambridge University Press, 1998), pp. 168–90.

Zinsser, Judith, 'Translating Newton's *Principia*: the Marquise du Châtelet's revisions and additions for a French audience', *Notes and Records of the Royal Society* 55 (2001), 227–45.

Zinsser, Judith, 'Entrepreneur of the "Republic of Letters": Emilie de Breteuil, Marquise Du Châtelet, and Bernard Mandeville's *Fable of the bees*', *French Historical Studies* 25 (2002), 595–624.

INDEX